Caribbean Primary
Mathematics

6th edition

Contributors

Jonella Giffard
Hyacinth Dorleon
Melka Daniel
Martiniana Smith
Troy Nestor
Lydon Richardson
Rachel Mason
Eugenia Charles
Sharon Henry-Phillip
Glenroy Phillip
Jeffrey Blaize
Clyde Fitzpatrick
Reynold Francis
Wilma Alexander
Rodney Julien
C. Ellsworth Diamond
Shara Quinn

HODDER
EDUCATION
AN HACHETTE UK COMPANY

Contents

How to use this book

This Student's Book covers all the key strands and requirements of the OECS and regional curriculum for Level 1. The book provides both learning notes and activities to help and encourage students to meet the learning outcomes for the level.

The content is arranged in topics, which correspond to the strands in the syllabus. So, for example, Topics 3 and 7 relate to the strand of Computation, while Topics 8, 10 and 12 relate to the various aspects that need to be covered in the strand, Measurement.

Topic 1, **Getting ready**, is a revision topic that allows you to do a baseline assessment of key skills and concepts covered in the previous level. You may need to revise some of these concepts if students are uncertain or struggle with them.

Each topic has an opening spread with the following features:

Questions to discuss with the students as they begin a new topic.

Eye-catching photographs and illustrations to stimulate the interest of the students.

Stimulating activities to get students thinking.

The questions and photographs relate to the specific units of each topic. You can let the students do all the activities (A, B and C, for example) as you start a topic, or you can do just the activities that relate to the unit you are about to start.

Each topic in this Student's Book is divided into units, which deal with different skills that need to be developed to meet the learning outcomes. For example, Topic 2 **Number sense (1)** is divided into four units: **A** Numbers to 20, **B** Counting to 100, **C** Sets and **D** Different ways of counting.

Teaching notes for each topic are provided. You will find these before the student material in each topic.

Each unit is structured in a similar way.

A student-friendly list of learning objectives.

Key word list. The words are bold and blue in the text.

Teaching text which explains concepts and provides examples.

Graded activities for consolidation.

Problem-solving activities.

Important symbols are highlighted.

A review activity.

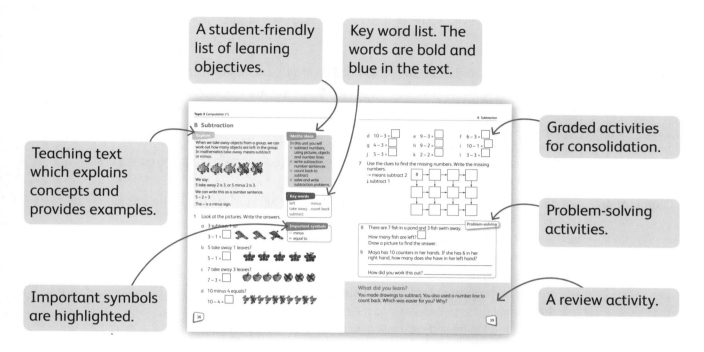

As you work through the units you will find a range of different types of activities and tasks including practical investigations, problem-solving strategies, projects and challenge questions. These features are clearly marked in the book so you know what you are dealing with.

The topics end with a review section that provides:

* A *summary activity* to help students consolidate and reflect on what they've learnt.
* *Think, talk, write …* activities which encourage students to share ideas, clarify their thinking and develop their maths dictionaries or write in their journals.
* A *quick check* revision exercise which includes questions from all the units.

There are three tests provided in the Student's Book to allow for on-going assessment and to prepare students for formal testing at all levels. Test 1 covers work from Topics 1–4, Test 2 covers work from Topics 5–9 and Test 3 covers work from Topics 10–13.

Teaching notes for Topic 1: Getting ready

This topic is designed to revisit in a fun way some of the key skills that the students should already have acquired. It is not a comprehensive review of the work done in kindergarten.

You can use the activities in this topic at the beginning of the year as part of your baseline assessment of the students or you can use the relevant pages as you begin a new topic with the class. If the students cannot do any of these activities, you will need to revisit some of the concepts and skills taught in kindergarten.

	Check students can:
Number concepts	* relate numbers 0–12 to groups of real objects in different arrangements. For additional practice, arrange the same number of real objects in different ways and let students count these. * compare groups of objects (has more/less). Give them real objects to make groups (sets). * read the number names from zero to twelve. Write the number names on posters or word cards and display them in the class so that students see them every day. * count forwards (to 50) and backwards (from 10) in sequence. Play counting games or sing counting songs if students need more practice.
Computation	* count the number of objects in two sets and reach a total (up to 9). Give students lots of additional practice using real objects and pictures. * subtract one number from another using real objects or pictures and separate objects by taking away from a set. * understand and write simple number sentences to show addition and subtraction. For additional practice, make prompt cards with numbers 1–10 and the signs + and = and let students make up their own number sentences using the cards. * use simple addition and subtraction in their everyday lives. Let them play a shopping game in which they use coins to pay for items in a low number range.
Geometry	* describe basic shapes using words such as round, straight, flat and curved. Revise this vocabulary using real objects such as balls, rulers and string. * sort and classify some plane (2-D) and solid (3-D) shapes. Help students to see the difference between the shapes by asking questions such as, 'How many sides does it have? Are all the sides the same? Are the sides straight or round? Is the shape flat?' * name squares, rectangles, triangles, circles, cones and spheres. Make sure that the shapes are displayed in the class with their names so that students become familiar with them. * identify and make patterns with shapes. Let students make patterns with real objects.
Shape and space	* describe the relative positions of objects using words such as above, below, in, on, left, right, inside, outside, in front of and behind. Play games to reinforce the meanings of these position words.
Measurement	* distinguish between times of day (morning, afternoon, evening) and different days (today, tomorrow, yesterday). * name the days of the week in order. Make sure the names are displayed in the class so that the students learn to read these by sight. * tell the time (on the hour). Use model clocks to practice this and relate the times to times in the lives of the students. * describe length and distance using words such as long, short, near and far. Use real objects and distances to reinforce these words. * describe mass using words such as heavy and light. * identify containers that hold more or less than other containers.
Data and graphs	* collect and classify simple sets of data from their own environment. Give students practice with small, safe, real objects from their environment (such as stones, sticks, leaves and feathers). * represent data by using blocks or pictures. Use small pictures or blocks to make very simple graphs with the class.

Getting ready

Numbers

1 How many in each set?
Write the numbers.
Circle the set with the most fruit.

a **b** **c**

2 Draw the correct number of candles on each cake.

    a three 3   **b** seven 7   **c** four 4

3 Fill in the missing numbers.

Adding and subtracting

1 How many altogether? Write the number.

   a  = ☐

   b  = ☐

2 Draw one more. How many are there now? Write the number.

   a 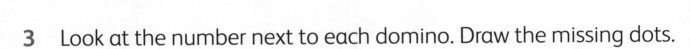   ☐

   b    ☐

3 Look at the number next to each domino. Draw the missing dots.

   a  = 6   b = 8

   c = 4   d = 7

4 Take away. How many are left? Complete the number sentences.

   a 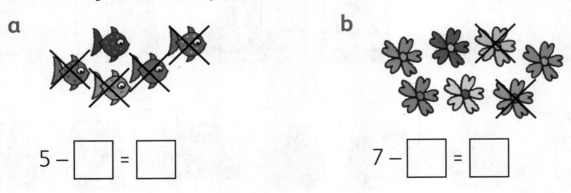   b

      5 – ☐ = ☐   7 – ☐ = ☐

Shapes and patterns

1 Match the pictures with the names of the shapes.

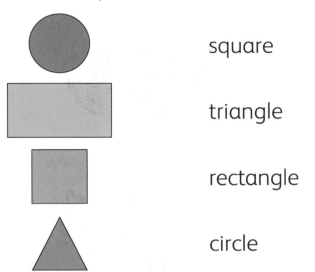

square

triangle

rectangle

circle

2 How many are there?

3 Complete the pattern.

Where is it?

1 Where is the bird? Circle the correct word.

a

on / in the branch

b

on / in the basket

c

in front of / inside the nest

d

above / below the tree

2 Draw and colour the shapes in each position.

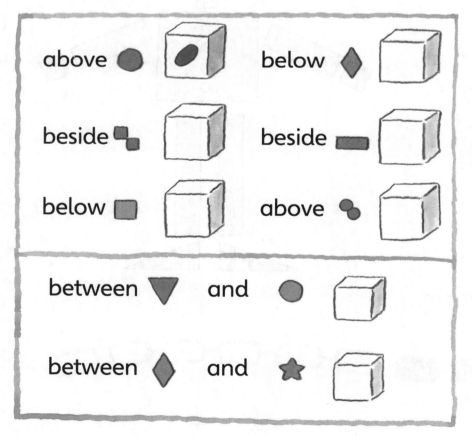

Measuring

1   What is the time?

_____ o'clock   _____ o'clock   _____ o'clock

2   Match the pictures and the times.

afternoon   morning   evening

3   Circle the object that is heavier.

4   Circle the container that holds more.

5   Say the names of the days of the week. Point to the names as you say them. Say what you do on each day.

Sunday	Monday	Tuesday	Wednesday	Thursday	Friday	Saturday

Sorting

1 Circle objects to make smaller groups. Describe the groups.

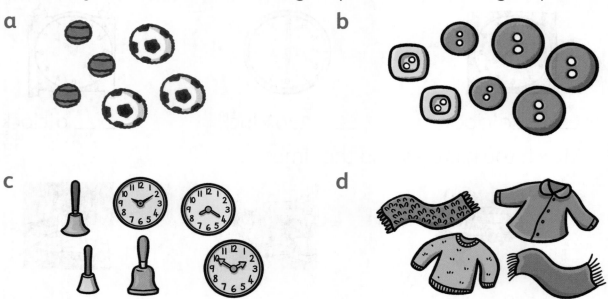

a

b

c

d

2 Count the socks in each colour. Draw the socks on the pictograph.

Colour	Number of socks					
pink						
blue						
orange						
green						

Key: ⌐ means one sock

Teaching notes for Topic 2: Number sense (1)

Overview

In this topic students will focus on counting forwards, backwards, on from a given number, and in groups (skip counting) in a higher number range. They will also represent numbers by making sets, writing numerals and saying number names.

A Numbers to 20	* Make sure students understand cardinality by matching number names to sets of real objects. They should handle and count objects to develop their skills. * Display numbers, representations of that amount and number names in the classroom to reinforce what you are teaching. * A numeral is a shorthand ways of writing a number name, but it doesn't show how many there are – students will only develop that knowledge by experience.
B Counting to 100	* In order to count on, students must understand where to start. * The last number we count tells us how many there are of what we are counting. * To compare numbers, students need to decide which is greater or smaller. * Ordering numbers involves putting numbers into ascending or descending order. * It is useful to print 1–100 blocks for each student so they can refer to this as they need to.
C Sets	* Making sets of real objects allows students to compare amounts to find, for example, that sets are equal, or that set A has one more than set B. * Sets also allow students to build number concepts by experiencing that two objects plus one more, makes three and so on. Similarly, that if you take away two objects from a set of five you are left with three.
D Different ways of counting	* Besides counting forwards and backwards, students also need to be able to count in groups, for example, count pairs in twos, or count fingers and toes in fives or tens. This is an important skill that can be practiced on a daily basis using real objects, for example, count the students in twos as they enter the classroom; count in fives (for example) as you hand out books to each group.

Mathematical vocabulary

Learning to use the correct mathematical vocabulary is an important skill to develop in early years. Students are encouraged to make their own mathematics dictionaries for this purpose. Each student will need his or her own notebook to use as a dictionary.

Notes for home

Encourage your child to find numbers in the environment and to say the number names.

Count items at home to reinforce the idea of a number of objects. For example, count out five spoons one by one. Once you've done that, ask, 'How many are there?' Add one more spoon and ask, 'How many are there now?' As your child learns more about numbers and counting, he or she will be able to say six without counting the spoons over again.

1234

Topic 2 Number sense (1)

A Numbers to 20 *(pages 12–17)*

Guess how many conch shells there are. Now count the shells. How many do you count?

✳ Count to 20. Say each number.
1 2 3 4 5 6 7 8 9 10
11 12 13 14 15 16
17 18 19 20

B Counting to 100 *(page 18)*

How many beads are there in this picture? Can you count them?

✳ What is the biggest number you know? What is the number before it? What is the number after it?

1 2 3 4

C **Sets** *(pages 19–22)*

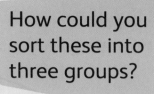

How could you sort these into three groups?

Describe each set.

A

B

C

* What is in the set?
* How many objects in the set?

D

How many socks on the washing line? How can you count the socks on this line quickly?

Different ways of counting *(pages 23–24)*

* How can you use these to count?

11

A Numbers to 20

Explain

How many crayons are there in this box?

We count the crayons to find out how many there are.

We count one by one: 1, 2, 3, 4, 5, 6.

The last **number** that you count tells you how many there are.

There are 6 crayons in this box.

Maths ideas

In this unit you will
* practice counting to 20
* count **backwards** from 10
* read and write **number names** from 1 to 20.

Key words

backwards
number names
number

1 Write the numbers.

zero	0	0	0	_____
one	1	1	1	_____
two	2	2	2	_____
three	3	3	3	_____
four	4	4	4	_____
five	5	5	5	_____
six	6	6	6	_____
seven	7	7	7	_____
eight	8	8	8	_____
nine	9	9	9	_____
ten	10	10	10	_____

2 Count backwards from 10. Say each number.

10   9   8   7   6   5   4   3   2   1

3 **a** How many fish do you see? Write the number.

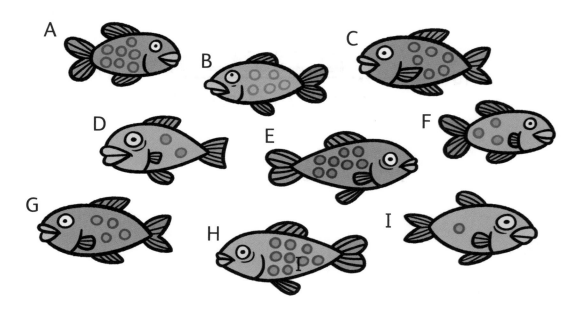

b How many spots on each fish? Write the numbers.

A _____   B _____   C _____

D _____   E _____   F _____

G _____   H _____   I _____

4 Count the dots. Write how many there are.

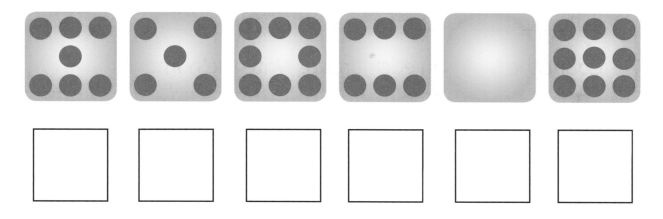

Challenge

5   Join the dots in the order of the numbers.

Explain

Read the number names.
Point to each number as you read the number name below it.

1	2	3	4	5	6	7	8	9	10
one	two	three	four	five	six	seven	eight	nine	ten

6   Complete the number names in order from 1 to 10.

_____ two   _____ four

five _____ seven   eight   _____

_____

7   Count the blocks in each group. Write the numbers.

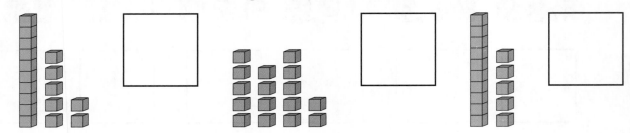

8 Draw a row of dots to show each number.

12:

15:

20:

16:

9 Fill in the missing numbers.

1			4	5			8		
		13	14		16		18		

10 Write these numbers in order from the smallest to the greatest.

a 6, 9, 3, 12   _____

b 18, 8, 10, 1   _____

c 20, 0, 10, 15, 5   _____

Problem-solving

11 Janice counted some shells. There are more than 14, but less than 17. The answer is not 15. What is the answer?

Investigate

12 How can you group these dots to make them easy to count? Draw the dots in groups.

15

> **Explain**
>
> Read the number names.
> Point to each number as you read the number name below it.
>
> | 11 | 12 | 13 | 14 | 15 |
> |----|----|----|----|----|
> | eleven | twelve | thirteen | fourteen | fifteen |
> | **16** | **17** | 18 | **19** | 20 |
> | sixteen | seventeen | eighteen | nineteen | twenty |

13 Write the number names. Then draw dots to show the numbers.

15	18
_____	_____

14 Write the number for each number name. Draw lines to match the numbers to the sets of dots.

eleven	twenty	thirteen	fifteen	nineteen

Project

15 Make your own mathematics dictionary.
You will need a small exercise book.
Write the letters of the alphabet in order, one letter at the top of each page.

As you learn new words in mathematics you can add words to your mathematics dictionary.

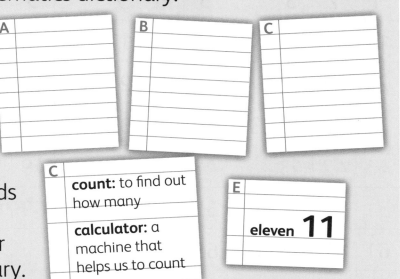

count: to find out how many

calculator: a machine that helps us to count

eleven 11

What did you learn?

1 Fill in the missing numbers.

a   1, 2, 3, ☐, ☐, 6

b   10, 9, 8, ☐, ☐

c   6, ☐, 8, 9, ☐, 11

d   10, ☐, 8, 7, ☐

2 Write the numbers.

a   The number after 10 is ☐

b   The number before 20 is ☐

c   The number between 13 and 15 is ☐

3 Complete the table of numbers and number words.

7	
	nine
	eleven

	twelve
	fourteen
19	

B Counting to 100

1	2	3		5	6	7		9	10
11	12	13		15	16	17	18		20
21	22	23		25	26	27	28	29	30
31	32	33	34			37	38	39	40
41	42	43		45	46	47	48	49	50
51	52		54	55	56	57	58	59	60
61	62	63	64	65		67	68		
71			74	75	76	77	78	79	80
81	82	83	84	85	86		88	89	90
91	92	93	94	95	96	97			100

Maths ideas

In this unit you will
* **count** to 100.

Key words

count
before
after

1   Read the numbers aloud. Work out what the missing numbers are and fill them in.

2   Circle the number **before** each of these numbers:
17, 39, 50, 87, 100.

3   Underline the number **after** each of these numbers:
23, 44, 78, 62, 99.

Challenge

4   Circle the number that does not belong in each set.

   a 21   22   23   35   25   **b** 34   35   63   37   38

What did you learn?

Work in pairs. Take turns to choose a number from 1 to 100. Give your partner clues.

It's before 60 and after 58.

C Sets

A set is a collection of **objects**.

The objects in the **set** can be the **same**.

This is a set of ice-creams.
There are 8 ice-creams in the set.

The objects in a set can be **different**.

There are 4 objects in this set. Each object is different.

A set with no objects is an **empty** set.

Maths ideas

In this unit you will
* count and name objects in sets
* make and draw objects in sets.

Key words

objects	different
set	empty
same	

1   What are the objects in each set? How many objects in each set? Write the numbers.

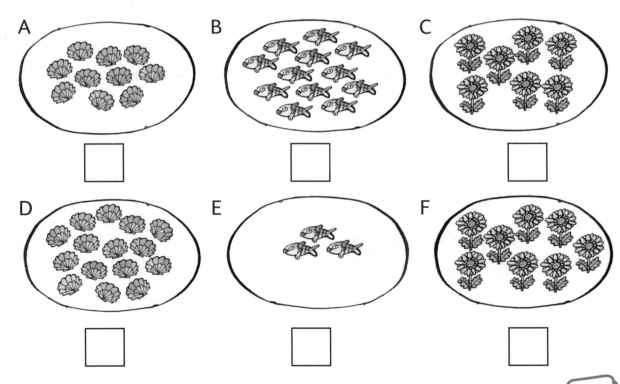

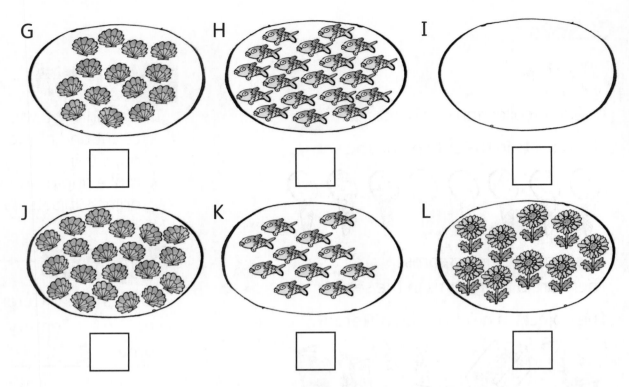

G ▢   H ▢   I ▢

J ▢   K ▢   L ▢

2   How many objects in each set? Write the number name under each set.

six   ten   twenty   eleven

a

b

_____   _____

c   d

_____   _____

3   Draw more objects to make sets.

a set of 5	a set of 9
a set of 12	a set of 8
a set of 20	a set of 15

4   Make these sets and then draw them.

   a   a set of 17 counters   **b**   a set of 12 beans

c a set with six different objects **d** a set of two buttons

e an empty set **f** a set of ten crayons

Investigate

5 Find three sets in your classroom. Write the names of each set.

6 How many objects are there in each set?

Set of objects	Number of objects in the set

What did you learn?

True or false? Write T or F.

1 A set is a collection of objects.

2 We can count the objects.

3 All sets have the same number of objects.

4 An empty set has no objects.

D Different ways of counting

Explain

We can **skip count** in groups to find out how many objects there are.

Counting the legs of these birds in 2s, is quicker than counting in 1s.

2   4   6   8   10

Maths ideas

In this unit you will
* count in different ways
* skip count in 2s
* use a **calculator** to count.

Key words

skip count   keys
calculator

1   Count the legs in 2s. Write how many legs there are.

a 

b 

2   Colour in every second number. Say the numbers aloud.

1	2	3	4	5	6
7	8	9	10	11	12
13	14	15	16	17	18
19	20	21	22	23	24
25	26	27	28	29	30
31	32	33	34	35	36
37	38	39	40	41	42
43	44	45	46	47	48
49	50				

3   Work in pairs. How many ways can you use to count the number of days in a week?

Did you use your fingers?   _____

Did you use counters?   _____

Did you count in your head?   _____

Which was the easiest way?   _____

4   Work in pairs. Count aloud in 10s from 10 to 100. Then fill in the numbers.

10	20	30		50					100

Explain

We can use calculators to count forwards and backwards.
We use the **keys** + and = to count forwards.
We use the keys − and = to count backwards.

To count in 2s
Press 0 + 2 = and keep on pressing = until you get to the number you want.

5   Use your calculator to answer these questions. Write your answers.

a   How old will you be in two years' time? _____

b   How many pencils do you have in your pencil case? _____

What did you learn?

How many ears are there? Count in 2s. Write how many.

1

2

Topic 2 Review

Key ideas and concepts

* Counting tells you the number of objects in a group. The last number that you count tells you how many there are.

* **12** is a number, which we call a numeral, and **twelve** is a number name. All numerals have number names.

* A **set** is a group or collection of objects.

* We can count quickly by skip counting in 2s and 10s.

* We can count forwards and backwards.

* We can use calculators to count.

Think, talk, write ...

1 Add these words to your mathematics dictionary. Draw pictures to show what each word means.

   a set

   b number name

   c number

2 Play a game in groups. Write the numbers from 0 to 50 on cards or pieces of paper. Place them face down in the middle of the group. Each player has a turn to pick up a card and give clues about the number. The player must not show the number or say what it is. The player who guesses correctly has the next turn.

   * What kind of clues can you give?

Quick check

1 Count the objects in each set and then write the numbers.

a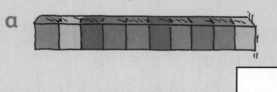

b

2 Fill in the missing numbers.

a 6, 7, ☐, 9, ☐

b 16, ☐, 18, ☐, ☐

c 7, ☐, 5, ☐, 3

d 31, ☐, 33, 34, 35

3 Draw the sets.

a a set of 23 buttons

b a set of 17 apples

4 Match the numbers and the number names.

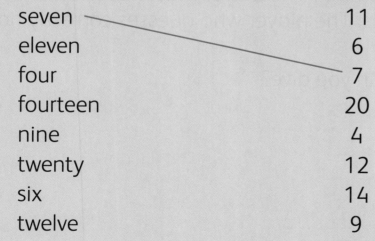

seven   11

eleven   6

four   7

fourteen   20

nine   4

twenty   12

six   14

twelve   9

26

Teaching notes for Topic 3: Computation (1)

Overview

In this topic students will focus on learning to add and subtract numbers up to 20 in different ways. They will learn number facts and learn to represent addition and subtraction in number sentences.

A Addition	* Using actual objects can help students work out answers in addition and subtraction. Have a collection of objects from inside or outside the classroom that are easily available and safe to use, such as buttons and small stones.
	* The students need to understand that we combine numbers to add them.
	* The total of an addition is called the sum of the numbers.
	* There are different methods of adding numbers. For example, we can use real objects and then count them, or we can use a number line and then count on to add numbers.
	* We can write number sentences to represent addition. We use the symbols + and = in these number sentences.
B Subtraction	* When we subtract, we find the difference between numbers.
	* To teach subtraction, begin with groups of real objects and let the students take objects away from a group.
	* We can write number sentences to represent subtraction. We use the symbols – and = in these number sentences.
C More about addition and subtraction	* Students need to learn addition and subtraction facts to 20 by memory so that they can add numbers quickly and easily in their heads.
	* Students can use addition and subtraction facts to work out (derive) other facts. If they know one fact, they can use it to work out the others in the number family.
	* A set of related facts using the same numbers is called a number family.
D Problem-solving	* Students need to learn to look for clues that tell them what type of calculation they need to do. These clues include the symbols + and – as well as words such as in total, altogether, double, take away and subtract.

Mathematical vocabulary

Students need to learn to use precise vocabulary such as **add**, **join together**, **take away**, **subtract**, **plus** and **minus** to describe addition and subtraction.

Notes for home

Your child is learning how to add and subtract. At this stage they are working with real objects to develop a good understanding of these two operations. This operation stage is important. You can help them by putting items together and taking some items away from a group. Students will learn more formal methods of working when they are ready.

1234

3 Computation (1)

A Addition *(pages 30–35)*

  and   is | 5 |

> Jo drew this picture in her mathematics book. What do you think it means?

How many in each group? How many altogether? Write the answer.

 and  is | |

B Subtraction *(pages 36–39)*

> Hold up 10 fingers. Fold away 5. How many fingers are left?

Four boats come into dock. Then 2 sail away early. How many are left?

 | |

1234

C **More about addition and subtraction** (pages 40–43)

Look at the two pictures. What has happened? How many bananas are left in the bowl? Do you have to add or subtract (take away) to get the answer?

You catch seven fish and give three of the fish to your friend. How many fish will you have left? Draw a picture to show this.

D

If Tyrone jumps forwards two blocks, on which number will he land? If he jumps backwards one block, where will he land?

Problem-solving (pages 44–45)

What do you have to do to solve these problems? Write + (add) or − (subtract). How do you know?

a   Pete caught three fish yesterday. He caught two fish today. How many did he catch altogether?

b   Nikki has six oranges. She gives one to her friend, Tara. How many does she have left?

A Addition

Explain

When we **add** objects, we find out how many there are **altogether**.

 +

Two **plus** two is four altogether.
Two plus two makes four.

 +

Two plus one is three altogether.
Two plus one makes three.

 + +

Two plus three plus one is six altogether.
Two plus three plus one makes six.

Maths ideas

In this unit you will
* add numbers, using pictures, objects and number lines
* write **addition** number sentences
* solve and write addition problems.

Key words

add   plus
altogether   equals
addition

Important symbols

+ plus
= equal to

1   Add the objects.

a

☐ plus ☐ is ☐.

b

☐ plus ☐ is ☐.

c

☐ plus ☐ is ☐.

d

☐ plus ☐ is ☐.

e ♥♥♥♥ + ♥♥ + ♥

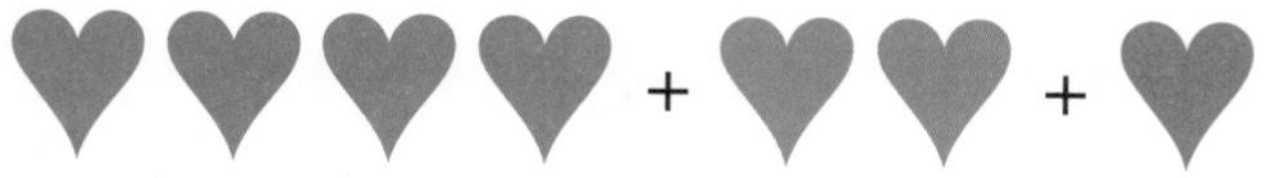

☐ plus ☐ plus ☐ is ☐.

f ◆◆◆◆◆◆◆ + ◆ + ◆

☐ plus ☐ plus ☐ is ☐.

2 Which of the groups above has the most objects? Circle the letter.

a   b   c   d   e   f

3 Draw dots to show the groups. Add the two groups.

●●●●	●●	●●●●●●
 4	 2	 makes _____ altogether
1	3	makes _____ altogether
5	2	makes _____ altogether
9	1	makes _____ altogether
2	11	makes _____ altogether
8	6	makes _____ altogether
10	5	makes _____ altogether

31

Explain

We can **count on** to find out how many objects there are altogether. You can count on using your fingers or you can count on using a number line.

What is 2 and 3 altogether?

Start with the first number, 2.
Count on for 3.
The answer is 5.

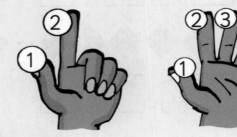

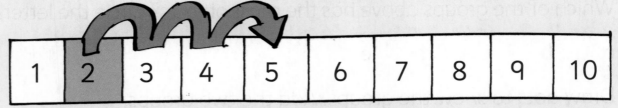

Start on 2. Count on for 3. The answer is 5.

4   Count on to add these numbers. Use your fingers.

1 plus 3 more is ☐   3 plus 3 more is ☐

4 plus 2 more is ☐   6 plus 3 more is ☐

5   Fill in the missing numbers on the number lines. Count on to add.

a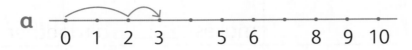

2 plus 1 makes ☐

b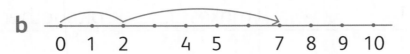

2 plus 5 makes ☐

c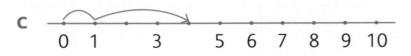

1 plus 3 makes ☐

6 Count on to add these numbers. Use the number line.

0 1 2 3 4 5 6 7 8 9 10 11 12 13 14 15 16 17 18 19 20

a 12 plus 6 makes ☐

b 15 plus 5 makes ☐

c 11 plus 2 makes ☐

d 19 plus 1 makes ☐

Explain

When we add numbers to find out how many there are altogether, we are doing **addition**. If we add 2 and 3, we make 5 altogether.
We can write what we do as **number sentences**, like this:
2 + 3 = 5
+ means plus
= means equals
So, we can say: two **plus** three **equals** five.

7 Complete the number sentences.

a

5 + 2 = 7

b

3 + 1 = ☐

c

2 + 3 = ☐

d

3 + 4 = ☐

e

1 + 4 = ☐

f

5 + 1 = ☐

8 Write number sentences for each of the groups of dots.

   a   **b**

_____   _____

9 Write number sentences for these dominoes.

a 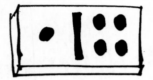   **b**    **c**

_____   _____   _____

Explain

We can add numbers in a different order and get the same answer.

$4 + 1 = 5$ $10 + 6 = 16$

$1 + 4 = 5$ $6 + 10 = 16$

10 Complete these number sentences.

    a $4 + 3 =$ _____ $3 + 4 =$ _____

    b $5 + 2 =$ _____ $2 + 5 =$ _____

    c $9 +$ _____ $= 10$ $1 + 9 =$ _____

    d _____ $+ 8 =$ _____ $2 +$ _____ $=$ _____

    e $10 + 5 =$ _____ $5 +$ _____ $=$ _____

    f $2 +$ _____ $= 8$ $6 +$ _____ $=$ _____

    g _____ $+ 5 = 9$ $4 +$ _____ $=$ _____

11 Add these numbers. Complete the number sentences.

a 2 + 3 + 4 = ☐

b 6 + 1 + 1 = ☐

c 10 + 5 + 5 = ☐

d 8 + 3 + 4 = ☐

e 11 + 3 + 3 = ☐

f 1 + 9 + 1 = ☐

Problem-solving

12 Work out the answer to this problem. You can use real objects, your fingers or a number line to work this out. You could also use a calculator to work out or check your answer.

Molly picked up 6 shells on Monday, 8 shells on Tuesday and 11 shells on Thursday. How many shells did she have?

☐

13 There are 5 people in Terri's family. One day 3 visitors came for supper. Terri put out nine plates and nine forks. What did she do wrong?

_____

What did you learn?

Complete the number sentences.

1 6 + ☐ = 10

2 9 + ☐ = 12

3 ☐ + 3 = 10

4 ☐ + 7 = 14

5 4 + 6 + 4 = ☐

6 1 + 10 + 3 = ☐

B Subtraction

When we take away objects from a group, we can work out how many objects are **left** in the group. In mathematics **take away** means **subtract** or **minus**.

We say:
5 take away 2 is 3, or 5 minus 2 is 3.

We can write this as a number sentence.
5 − 2 = 3

The − is a minus sign.

Maths ideas

In this unit you will
* subtract numbers, using pictures, objects and number lines
* write subtraction number sentences
* count back to subtract
* solve and write subtraction problems.

Key words

left	minus
take away	count back
subtract	

1 Look at the pictures. Write the answers.

Important symbols

− minus
= equal to

a 3 subtract 1 is?

3 − 1 = ☐

b 5 take away 1 leaves?

5 − 1 = ☐

c 7 take away 3 leaves?

7 − 3 = ☐

d 10 minus 4 equals?

10 − 4 = ☐

2 Write a number sentence for each picture.

a

6 – _____ = _____

b

8 – _____ = _____

c

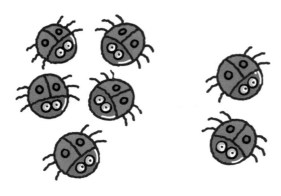

7 – _____ = _____

d

10 – _____ = _____

3 Read the number sentence. Cross out the shapes to take them away. Write how many are left.

a

8 – 3 = ☐

b

10 – 2 = ☐

c

9 – 1 = ☐

d

8 – 4 = ☐

4 Write your own number sentence for each picture.

a  b c

_____ _____ _____

5 Write the answers to complete these number sentences.

a ⬤⬤⬤⬤⬤⬤⬤⬤⬤⬤ $10 - 6 =$ ☐

b ⬤⬤⬤⬤⬤⬤⬤⬤⬤⬤⬤⬤ $12 - 6 =$ ☐

c ⬤⬤⬤⬤⬤⬤⬤⬤⬤⬤⬤⬤⬤⬤⬤ $15 - 5 =$ ☐

d ⬤⬤⬤⬤⬤⬤⬤⬤⬤⬤⬤⬤⬤⬤⬤⬤⬤ $17 - 8 =$ ☐

e ⬤⬤⬤⬤⬤⬤⬤⬤⬤⬤⬤⬤⬤ $13 - 7 =$ ☐

f ⬤⬤⬤⬤⬤⬤⬤⬤⬤⬤⬤⬤⬤⬤⬤⬤ $16 - 10 =$ ☐

Explain

You can **count back** on a number line to subtract.

$8 - 2 =$ ☐

Start at 8.
Count 2 places back to subtract 2.
You land on 6. That is the answer.

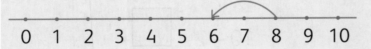

0 1 2 3 4 5 6 7 8 9 10 $8 - 2 = 6$

6 Use the number line to count back. Write the answers.

0 1 2 3 4 5 6 7 8 9 10

a $6 - 2 =$ ☐   b $7 - 3 =$ ☐   c $9 - 1 =$ ☐

d 10 − 3 = ☐

e 9 − 3 = ☐

f 6 − 3 = ☐

g 4 − 3 = ☐

h 9 − 2 = ☐

i 10 − 1 = ☐

j 5 − 3 = ☐

k 2 − 2 = ☐

l 3 − 3 = ☐

7 Use the clues to find the missing numbers. Write the missing numbers.

→ means subtract 2

↓ subtract 1

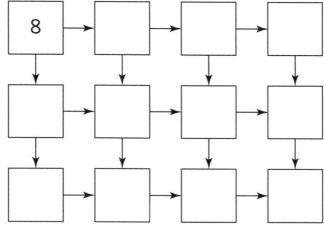

Problem-solving

8 There are 7 fish in a pond and 3 fish swim away.

How many fish are left? ☐

Draw a picture to find the answer.

9 Maya has 10 counters in her hands. If she has 6 in her right hand, how many does she have in her left hand?

_____

How did you work this out? _____

What did you learn?

You made drawings to subtract. You also used a number line to count back. Which was easier for you? Why?

C More about addition and subtraction

There are different ways to make up the number 10.

1 + 9 = 10	9 + 1 = 10
2 + 8 = 10	8 + 2 = 10
3 + 7 = 10	7 + 3 = 10
4 + 6 = 10	6 + 4 = 10
5 + 5 = 10	

If you learn these addition facts you will be able to add numbers quickly in your head.

Maths ideas

In this unit you will
* learn about **number facts** for addition and subtraction
* add numbers in your head.

Key words

number facts
double

Important symbols

+ plus
− minus
= equal to

1   Look at each group of dots. How many more dots do you need to make groups of 10?

* Do this in your head.

* Then draw the dots to see if you are correct.

a

b

c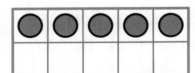

d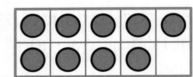

e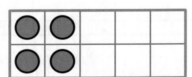

f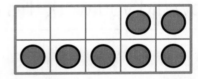

g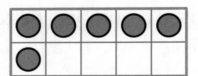

h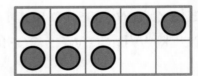

2 How quickly can you complete these number facts?

a 9 + ☐ = 10

b ☐ + 8 = 10

c ☐ + 2 = 10

d 7 + ☐ = 10

e ☐ + 4 = 10

f ☐ + 5 = 10

3 The students are making bead bangles. Each bangle needs 10 beads. Write a 10 fact for each bangle.

a

☐ + ☐ = 10

b

☐ + ☐ = 10

c

☐ + ☐ = 10

d

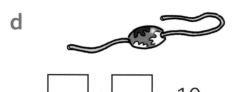

☐ + ☐ = 10

Explain

You can use addition facts to work out matching subtraction facts.

2 + 4 = 6      5 + 3 = 8

So,   So,

6 − 2 = 4   8 − 5 = 3

6 − 4 = 2   8 − 3 = 5

4   Complete the number sentences for each picture.

a

4 + 3 = 7

7 − 4 = ☐

3 + 4 = ☐

7 − 3 = ☐

b

4 + 2 = 6

6 − 4 = ☐

2 + 4 = ☐

6 − 2 = ☐

c

7 + 2 = 9

9 − 7 = ☐

2 + ☐ = 9

9 − ☐ = 7

d

☐ + ☐ = 8

8 − ☐ = ☐

☐ + ☐ = 8

☐ − 3 = 5

Explain

When we add two numbers that are the same, we **double** that number.

3 + 3 = 6   4 + 4 = 8

Here are some doubles facts you must know. If you know these facts, then you will be able to add numbers in your head more quickly. They will also help you do subtraction.

1 + 1 = 2   2 + 2 = 4   3 + 3 = 6   4 + 4 = 8   5 + 5 = 10

5   Double the numbers. Draw the missing dots on these dominoes. Write the number fact for each domino.

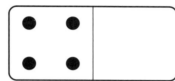

a

☐ + ☐ = ☐

b

☐ + ☐ = ☐

c

☐ + ☐ = ☐

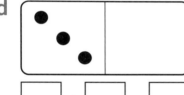

d

☐ + ☐ = ☐

e

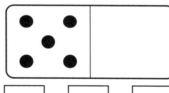

☐ + ☐ = ☐

f

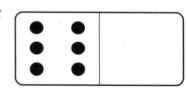

☐ + ☐ = ☐

What did you learn?

1   Double each of these numbers.

5 ☐   2 ☐   4 ☐   1 ☐

2   Complete the number sentences.

a   2 + 4 = ☐   ☐ − 4 = ☐

b   7 + 3 = ☐   ☐ − 7 = ☐

43

D Problem-solving

How do we know when to subtract or when to add?
We can look for symbols in number sentences:
+ means add
− means subtract

We can look for words in problems:
These words are clues that we need to **add**:
altogether, **in all**, **total**, **double**.
These words are clues that we need to **subtract**:
left, **take away**, **difference**.

Think about whether the answer will be more or
less than the number you started with.
* If the answer is less, you will subtract.
* If the answer is more, you will add.

Maths ideas

In this unit you will
* solve and create addition and subtraction problems.

Key words

add	subtract
altogether	left
in all	take away
total	difference
double	

Important symbols

+ plus
− minus
= equal to

1   Work with a partner. Decide whether you
     will add or subtract.
     Write a number sentence and work out
     the answer.

   a Amira found three shells. Nikki found
         nine shells. How many more shells does
         Nikki have than Amira?

   b There are five mangoes in a bag. How many mangoes in
         two bags?

   c Therese had 10¢. She lost 5¢. How much money does she
         have left?

2 Use any way that you know to solve these problems as quickly as possible.

a There are 6 eggs in a box. How many eggs are there altogether in two boxes?

☐ ☐ = ☐

b Deanna and Toriane have 9 beads altogether. If Deanna has five beads, how many does Toriane have?

☐ ☐ = ☐

c Micah catches 7 fish. Nathan catches 10 fish. How many more fish must Micah catch to have the same number as Nathan?

☐ ☐ = ☐

Challenge

3 Make up your own story problem for each number sentence.

a 4 + 6 = 10

_____

_____

b 9 − 4 = 5

_____

_____

What did you learn?

True or false? Write T or F.

1 Addition facts can help you do subtraction. ☐

2 If you double a number you add. ☐

3 Take away means to subtract. ☐

Topic 3 Review

Key ideas and concepts

* We can **count on** to find out how many objects there are altogether.
* When we add numbers we are doing **addition**.
* We can write addition number sentences both ways:
  2 + 3 = 5 OR 3 + 2 = 5
* When we **count back**, we **subtract**.
* **Take away** means **subtract** or **minus**.
* + means plus.
* − means minus or take away.
* = means equals.
* When we **double** a number, we add the same number.

Think, talk, write …

1   Which words in a story problem tell you whether you must add or subtract? Give four examples and say what they mean. Write these in your mathematics dictionary.

2   Give an example of a fact family. Use the numbers 3 and 4.

☐ + ☐ = ☐   ☐ + ☐ = ☐   ☐ − ☐ = ☐

Quick check

1   Write the answers.

a 9 − 6 = ☐   b 5 + 5 = ☐   c 10 − 5 = ☐

d 13 + 3 = ☐   e 20 − 6 = ☐   f 15 − 7 = ☐

2   Complete these number facts.

6 + ☐ = 10   ☐ + 4 = ☐   ☐ − 4 = 6

Teaching notes for Topic 4: Shape and space (1)

Overview

In this topic students will focus on recognising and describing these 2-D shapes: triangles, circles, rectangles and squares. They will also learn about position words.

A Shapes around us	* Recognising shapes helps students to sort and classify, which is an important concept in mathematics. * Two-dimensional or 2-D shapes are flat shapes (plane shapes). The two dimensions are length and width (breadth).
B Sorting and describing shapes	* Shapes can be named and identified by looking at the number and the length of sides and the number and types of corners (angles) the shapes have. Students can start by identifying shapes by the number of sides. * Circles are 2-D shapes with no straight sides or corners. * Triangles are 2-D shapes with 3 sides and 3 corners. * Squares have 4 equal sides and 4 corners (right angles). * Rectangles have 4 sides and 4 corners (right angles). The pairs of opposite sides are equal.
C Drawing and using shapes	* Recognising why things in our environment have certain shapes, helps us to draw, build and understand things around us. * Students can form patterns by using different shapes. This reinforces their ability to identify and sort the shapes.
D Where is it?	* We use position words (such as above, between, in, on) to describe where things are in relation to each other.

Mathematical vocabulary

Students need to learn vocabulary that they can use to describe shapes and learn to differentiate between them, for example: **size** (big, small) **sides**, **corners**. Along with position words, these are building blocks to understanding geometry.

Notes for home

Two-dimensional shapes are flat shapes, such as circles, squares, triangles and rectangles. You can find examples of these shapes in designs on fabrics, floor-tile patterns and pictures on playing cards. You can cut out shapes from coloured paper (or buy sticky shapes) and get your child to put them together them to make pictures. For example, they may make a rectangular house with a triangle for the roof and squares for the windows.

Shape and space (1)

A   **Shapes around us** *(pages 50–51)*

What shapes can you see in this piece of Native American fabric? How many of each shape do you see?

Look at these shapes. Colour in the shape that does not belong with the other shapes. Discuss why it does not belong.

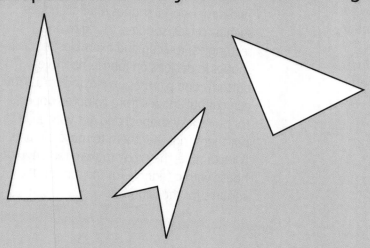

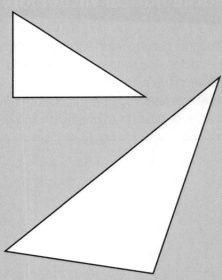

B Sorting and describing shapes *(pages 52–54)*

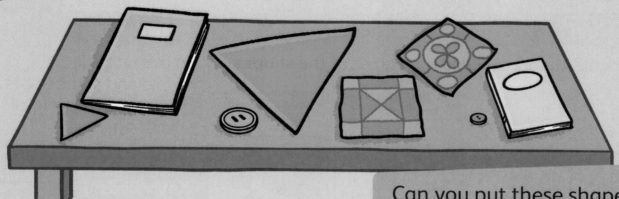

Can you put these shapes into groups? How will you decide on the groups?

C Drawing and using shapes
(pages 55–57)

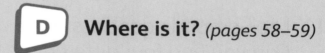

What shapes can you see in these patterns? What else helps to make the patterns?

Colour in the shapes to make a pattern.

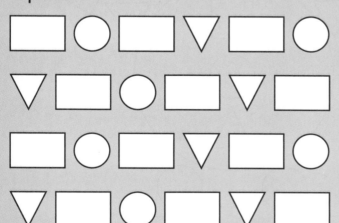

D Where is it? *(pages 58–59)*

Where is the green dot? How else can you describe its position?

Choose a mouse. Tell your partner where the mouse is. Let your partner point to the correct picture.

A Shapes around us

Explain

Look at the shapes. Read the names of the **shapes**.

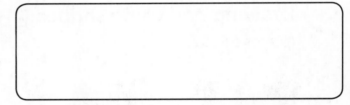

circles

squares

triangles

rectangles

Maths ideas

In this unit you will
* find and name circles, triangles, rectangles and squares.

Key words

shapes	triangles
circles	rectangles
squares	

1   Look around your classroom. Can you find four different shapes? Point to them. Draw two of the shapes here.

2   a   Colour in the picture.

b Count the shapes. Fill in the numbers.

Shape	Number
square	
circle	
triangle	
rectangle	

3 Read the labels. Draw and colour in the shapes.

a red circle	a blue triangle
a red square	a green rectangle

What did you learn?

Circle the correct name for each shape.

1   triangle   rectangle   2   rectangle   circle

3   triangle   square   4   circle   triangle

B Sorting and describing shapes

Explain

Some **shapes** have **corners** and straight **sides**.

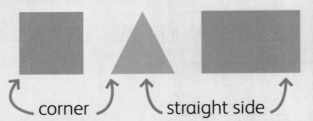

corner   straight side

* **Triangles** have 3 straight sides and 3 corners.
* **Squares** and **rectangles** have 4 straight sides and 4 corners.
* **Circles** have no corners and no straight sides.

Maths ideas

In this unit you will
* describe circles, triangles, rectangles and squares
* sort shapes into groups.

Key words

shapes	squares
corners	rectangles
sides	circles
triangles	size

1 Colour the shape on the right that is the same shape and **size** as the one on the left.

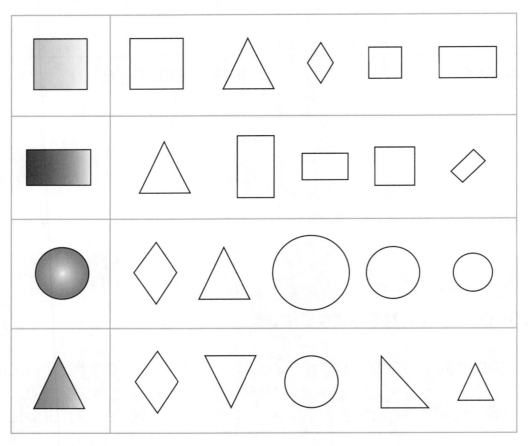

2 Fill in the correct shape names.

a A _____ has 3 sides and 3 corners.

b A _____ has no corners and no straight sides.

c Squares and _____ have _____ sides

and 4 _____ each.

3 Circle the corners. Colour the sides blue.

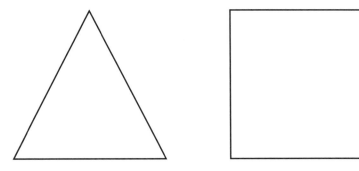

4 Read the labels. Draw the shapes.

I have 3 sides and 3 corners.	I have 0 corners and 0 straight sides.	I have 4 sides and 4 corners.

Explain

We can sort shapes by looking at:
* the colour
* the size
* the type of shape (the number of sides and corners).

5   Michael wants to divide these shapes into two groups or sets.

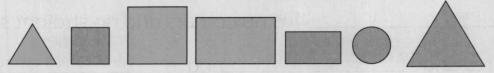

One group will have big shapes.
The other group will have blue shapes.
How should he divide them?
Draw the shapes in groups.

Share your ideas with your group.

What did you learn?

How have these shapes been sorted? Underline the correct answers.

1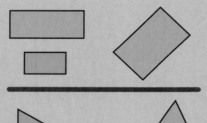

They are sorted into different
colour groups.

They are sorted into groups of
rectangles and triangles.

2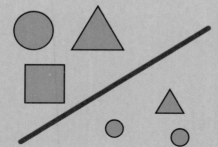

They are sorted into big and
small shapes.

They are sorted into different
colour groups.

C Drawing and using shapes

Explain

To make a pattern we need to repeat things.

We can repeat **shapes** to make a **pattern**.

We can repeat colours to make a pattern.

Maths ideas

In this unit you will
* draw circles, triangles, rectangles and squares
* use shapes to make patterns and pictures
* talk about patterns and pictures.

Key words

shapes   pattern

1   Colour in and draw. Complete the patterns.

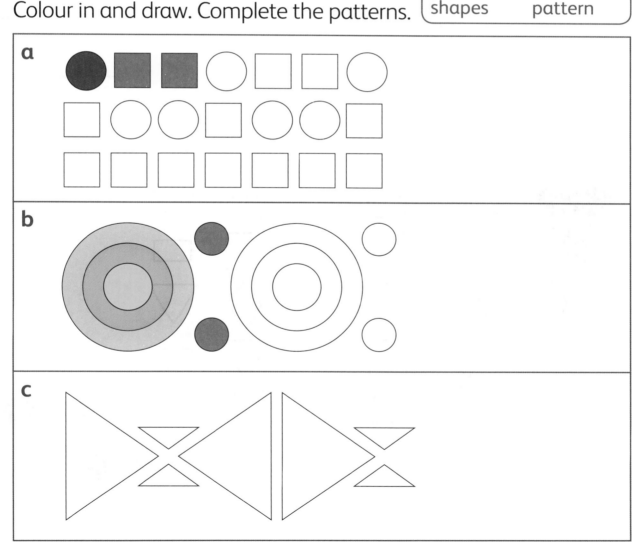

2 Draw beads to complete the necklaces. Colour the beads so that they make a pattern.

a

b

c

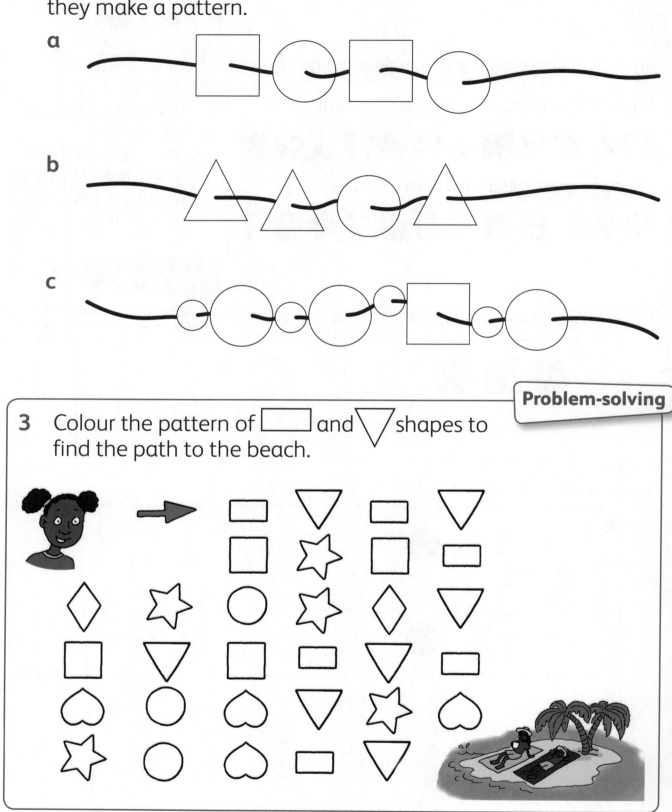

Problem-solving

3 Colour the pattern of ▭ and ▽ shapes to find the path to the beach.

4   Circle the mistake in each pattern. Draw the correct pattern.

a ■ ● ● ■ ● ● ●

b ▲ ● ▼ ● ▲ ● ▲

c ● ● ● ● ● ● ●

Investigate

5   Draw and colour in your own patterns. Fill the space.

What did you learn?

Draw what comes next in this pattern.

D Where is it?

Words such as **above**, **below** and **between** describe positions.

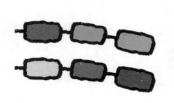

The blue bead is **above** the yellow bead.

The orange bead is **between** the blue bead and the pink bead.

The green bead is **below** the pink bead.

We can also use the words **on**, **inside** and **under**.

The bowl is **on** the table.

The ball is **inside** the box.

The box is **under** the table.

We can also use **top**, **middle** and **bottom** to describe position.

Maths ideas

In this unit you will
* use position words to describe where objects are.

Key words

above	under
below	top
between	middle
on	bottom
inside	

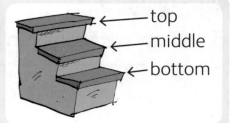

1   Circle the pictures that match the words.

on			
between			
under			
inside			
above			
below			

2 Colour the shapes in the table using different colours.

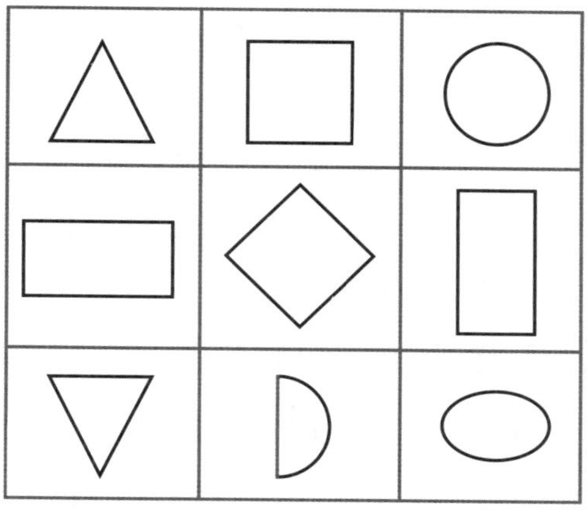

3 Find the correct shape in your table and draw it. Colour in the shape to match your table.

above 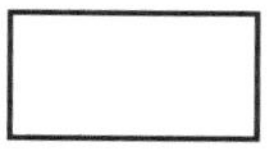 ▭	
below  ▯	
beside  △	
below ◻	
above  ◗	
between 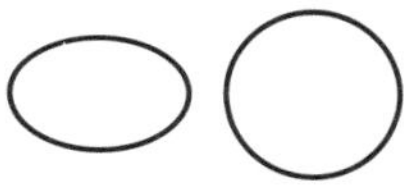 ⬭ ◯	

What did you learn?

Complete this picture. Follow the instructions.

1 Draw 4 circles on the top shelf.

2 Draw 3 squares on the bottom shelf. Colour the square in the middle red.

3 Draw 3 triangles and 3 rectangles in a pattern on the middle shelf.

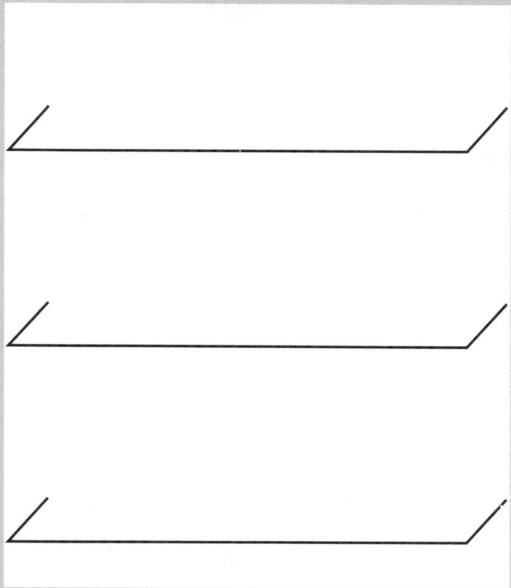

59

Topic 4 Review

Key ideas and concepts

✴ Some shapes, such as squares, rectangles and triangles, have **sides** and **corners**.

✴ **Circles** have no corners and no straight sides.

✴ **Triangles** have 3 straight sides and 3 corners.

✴ **Squares** and **rectangles** have 4 straight sides and 4 corners.

✴ We can **repeat** shapes and colours to make a pattern.

✴ Words such as **above**, **below**, **between**, **inside**, **under**, **on, top**, **middle** and **bottom** describe positions.

Think, talk, write …

1   Write 9 position words in your mathematics dictionary. Draw small pictures to remind yourself what each word means.

2   Think of three shapes that you see in everyday life. Share your ideas with your group.

3   How can you put these shapes into groups? Write two sentences in your journal if you have one.

Quick check

1   Circle the correct word under each picture.

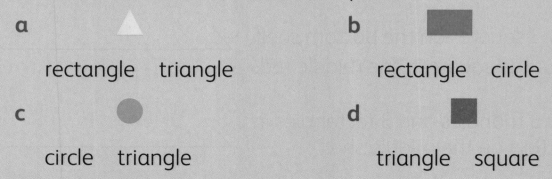

a   rectangle   triangle   b   rectangle   circle

c   circle   triangle   d   triangle   square

60

Test yourself (1)

1   What do you know about numbers? Listen to your teacher and colour in the faces.

I can:

count to 100 in different ways.	☺ always	☺ not yet
count backwards from 10.	☺ always	☺ not yet
read all the number names from 1 to 10.	☺ always	☺ not yet
write all the number names from 1 to 10.	☺ always	☺ not yet
read all the number names from 11 to 20.	☺ always	☺ not yet
write all the number names from 11 to 20.	☺ always	☺ not yet

2   Fill in the missing numbers.

a   11, 12 _____, 14 _____, _____, 17 _____, 19 _____

b   10, 9, _____, _____, 6 _____, _____, 3 _____, 1 0

3   Complete the table.

Shape	How many sides does this shape have?	What do we call this shape?	Your drawing
△	3		
☐		square	
◯		circle	

4 Add or subtract. Colour by number.

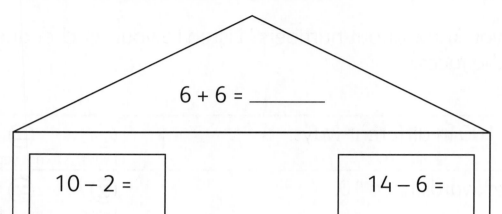

6 + 6 = _____

10 – 2 = _____

14 – 6 = _____

9 + 6 = _____

8 + 3 = _____

6 + 5 = _____

Key
red = 15
brown = 12
blue = 11
green = 8

5 Draw the picture.

　✱ Draw a big triangle in the middle of the space. Colour the triangle yellow.

　✱ Draw a small circle above the triangle.

　✱ Draw two big rectangles next to the triangle.

　✱ Draw four squares under the triangle.

Teaching notes for Topic 5: Number sense (2)

Overview

In this topic students will develop their understanding of numbers by comparing sets of numbers and learning about ordinal numbers. Then they will be introduced to place value and fractions.

A Comparing sets and numbers	* Make sure that students understand what makes up a set of things and then let them compare sets. Do this with real objects as you introduce the concepts equal to, more than and less than.
B Ordinal numbers	* The numbers we use for counting are called cardinal numbers. They tell us how many there are of something. * Ordinal numbers are used to describe the position of items. For example, the order in which runners finish in a race – first (1st), second (2nd) or third (3rd). * We can write ordinal numbers using numerals with the short forms -st, -nd, -rd or -th after them. Students should be able to recognise the words and the numerals.

C Place value	* Once students understand how to count and then how to add real objects, they can be introduced to the concept of place vlaue, which will allow them to add numbers in columns.
	* Place value is the value of a digit in a number.
	* The place of the digit in a number determines its value. For example:
	   – 13: value of 1 = one 10, value of 3 = 3 units
	   – 24: value of 2 = two 10s, value of 4 = 4 units.
	* Students will learn to write numbers as a sum of the values in each place. This is called expanded notation. For example, 25 = 20 + 5.

D Fractions	* Understanding basic fractions is an important building-block skill in mathematics as it helps students to understand other concepts, such as division. * Fractions are parts of a whole. A whole has two halves. One half of a whole is represented as $\frac{1}{2}$. * A whole has four quarters. One quarter is represented as $\frac{1}{4}$. * It is best to first teach students about fractions through pictures and by cutting up real objects into parts (fractions).

Mathematical vocabulary

Students need to use precise vocabulary to compare sets of objects (**equal to**, **more than**, **less than**), to name ordinal numbers (**first** and **second**, for example), to talk about place value (**tens**, **ones**) and to describe fractions (**whole**, **half**, **quarter**).

Notes for home

Help your child to understand simple fractions (half and quarter) at home. If you cut a cake or an apple or a pizza, first cut it in half and then into quarters. Say things like, 'This is the whole cake. Now let's cut it in half. Now let's cut it into quarters. How many pieces are there?' Try to cut slices that are equal in size.

1234

A Comparing sets and numbers *(pages 66–67)*

Which groups have the same number? Is there a pencil for each child?

Look at the two sets.

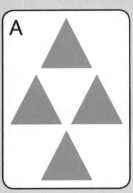 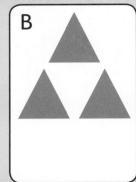

A   B

Which set has more objects?

A or B? ☐

Show what you can do to make the number of objects in each set the same.

B Ordinal numbers *(page 68)*

These are the winners of a race. Who came first? How do you know? Who came second? Who came third?

Ben   Teri   Nick

If you come second in a race, are you before or after the winner of the race?

Who was second in this race?

Who was third?

1234

C Place value (pages 69–70)

Both these groups of coins show 15 cents. How are the groups different?

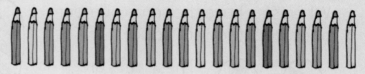

How many bundles of 10 can you make with these pencils?

Circle the bundles and write your answer.

How many pencils will you have left over?

How do you write the number of pencils?

D Fractions (pages 71–76)

You want to share your cookie with your friend. You give her half the cookie. How much is that? Can you show half on the picture?

Sandy bought a cake and shared it with some friends. They each got exactly the same. How many people shared the cake? What did Sandy do to share it out?

A Comparing sets and numbers

When sets have the same number of objects, they are **equal**.

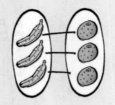

These **sets** are not equal.

The set of pencils has one **more than** the set of books.

The set of spoons has one **less than** the set of people.

Maths ideas

In this unit you will
* make and draw sets
* write the number of objects in a set
* compare sets and numbers using the symbols <, > and =.

Key words

equal

sets

more than

less than

1   Draw the sets. Write the numbers.

   a   Two sets of fish. The sets must be equal.

Important symbols

= equal to

> more than

< less than

   b   Two sets of flowers. One set has one more flower than the other set.

Explain

We can compare sets and numbers by using symbols.

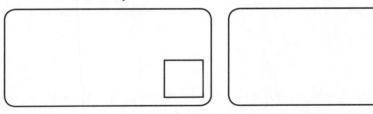
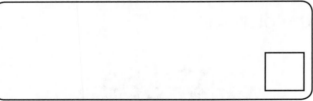

> means **more than**   7 > 6   14 > 13

< means **less than**   2 < 3   10 < 20

= means **equal to**   5 = 5   16 = 16

2 Fill in < or >.

a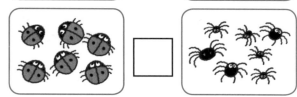

b

c

d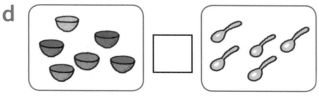

3 Make sets with one more or one less.

a [　　　　　　　　　] > ● ● ● ● ● ●

b [　　　　　　　　　] < ■ ■ ■ ■ ■ ■ ■ ■ ■ ■ ■

4 Fill in < or >.

a 8 ☐ 9　　　　b 19 ☐ 18　　　　c 6 ☐ 7

d 11 ☐ 10　　　e 15 ☐ 10　　　f 13 ☐ 17

Problem-solving

5 How could you make these sets equal? Draw pictures.

What did you learn?

Look at the pictures. Complete the sentences.

1 Set A is [　　　] set B.

2 a Set D is [　　　] than set C.

   b Set [　　] is > than set [　　].

A 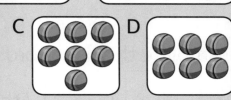 B

C

D

B Ordinal numbers

Explain

We use numbers such as 1, 3 and 5 to count.
We use numbers such as **1st, 2nd, 3rd** and **6th** to give the **position** of things. We call these **ordinal numbers**.

A   B   C   D   E   F   G   H   I   J

A is the first (1st) letter of the alphabet.
J is the **tenth** (10th) letter.

Maths ideas

In this unit you will
* learn about ordinal numbers such as 1st, 2nd and 3rd.

Key words

first (1st)   seventh
second   (7th)
  (2nd)   tenth
third (3rd)   (10th)
fourth   position
  (4th)   ordinal
fifth (5th)   numbers
sixth (6th)

1   Look at the picture and answer the questions.

1st

a   Which boat is 2nd? _____

b   What colour is the **5th** boat? _____

c   Which boat is behind the **4th** boat? _____

d   Which boat is in front of the 3rd boat? _____

2   Follow the instructions.

a   Colour the 2nd circle red.

b   Colour the 4th circle blue.

c   Colour the 1st circle yellow.

d   Write the positions of the other three circles.

_____ 2nd _____ 4th _____

What did you learn?

Complete this list of ordinal numbers.

1st ☐ ☐ 4th 5th ☐ **7th** ☐ ☐ ☐

C Place value

Explain

This is what 10 **ones** look like:

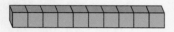

We can join these together to make 1 ten.

This is what 1 **ten** looks like:

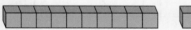

This is what 17 ones look like:

We can show this as 1 ten and 7 ones.

Maths ideas

In this unit you will
* make groups of **tens** and ones
* learn that we can write numbers as tens and ones
* learn how the position of a **digit** tells us the **value** of the digit.

Key words

ones	digit
tens	value

1   Circle the groups of ten. Write the number of tens and ones.

a

tens	ones

b

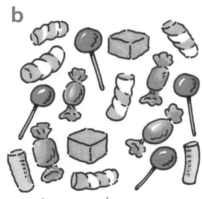

tens	ones

c

tens	ones

Problem-solving

2   Trevor has 23 mangoes. He sells them in packets of 10.

How many packets can he make from his 23 mangoes? ☐

How many loose mangoes will he have left? ☐

Explain

Numbers can have more than one **digit**. For example:

17

This number has **two digits**: 1 and 7

The **position** of a **digit** in a number tells us the **place value** of that digit.

17 = one ten and seven ones (10 + 7)

tens	ones
1	7

3 Say each number aloud.

tens	ones
2	5

tens	ones
3	3

tens	ones
1	9

tens	ones
2	0

4 How many tens are there in each of these numbers? Circle the digits that show the tens.

a 55   b 16   c 21   d 47

Challenge

5 A number has a 5 in the tens place and a 7 in the ones place.

What is the number? _____

Is the number bigger or smaller than 75? Circle < or >.

How do you know? _____.

What did you learn?

How many groups of tens and ones can you make?

1 ☆☆☆☆☆☆☆☆☆☆☆☆☆☆☆
☆☆☆☆☆☆☆☆☆☆☆☆☆☆☆

tens	ones

2

tens	ones

D Fractions

Explain

Halves

whole

This is a **whole** circle.

half   half

We can divide the circle into two **equal** parts. Each **part** is called one **half**. Two **halves** make one whole.

Maths ideas

In this unit you will
* find out what a whole and part of a whole are
* find out about halves and quarters (fourths)
* write and read fractions $\frac{1}{2}$ and $\frac{1}{4}$.

Key words

whole	quarter
equal	quarters
part	fourths
half	fraction
halves	

Important symbols

$=$ equal to
$>$ more than
$<$ less than

1 Colour in one half of each shape.

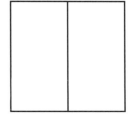
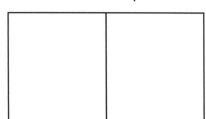

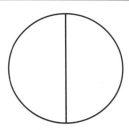

2 Tick the pictures that show halves.

a
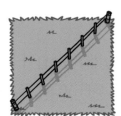

b
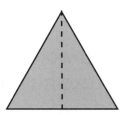

c

d

3   Draw a line on each shape to divide it into two equal halves.
    Colour in one half of each shape.

Investigate

4   Take a sheet of paper and fold it into two halves. Can you fold
    it in half in more than one way?

Problem-solving

5   Sandy cut a cake into two pieces. One piece
    was bigger than the other piece.
    Did she cut the cake into halves? _____
    How do you know? _____

6   Your friend wants to share his eight sweets with you. He says
    you can take half of the sweets. How many can you take?
    Circle the sweets.

Explain

Quarters (fourths)

whole

quarter

This is a whole circle.   We can divide the circle into four equal parts. Each part is called one **quarter**. Four **quarters** make one whole. Quarters are also called **fourths**.

7 Tick the shapes that have one quarter coloured in.

a 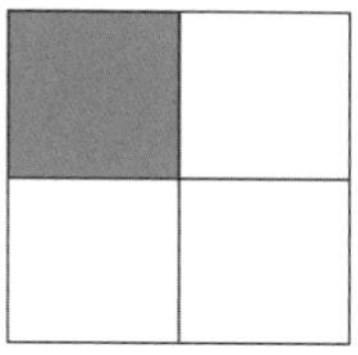   b 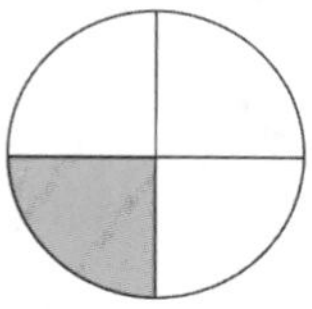   c 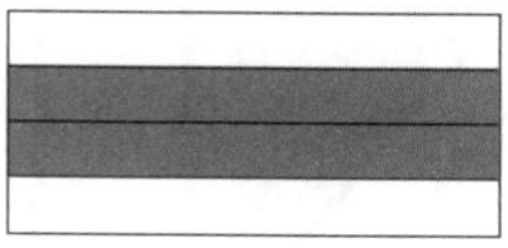   d

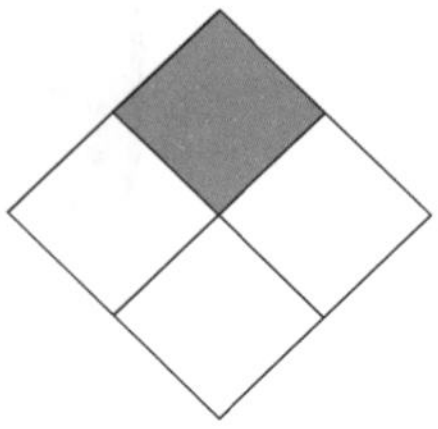

8 Colour in one quarter of each shape.

a 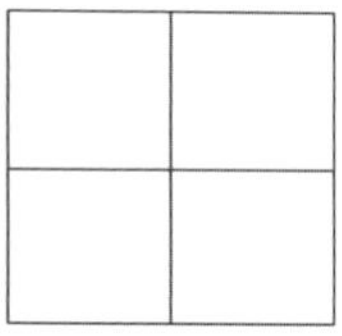   b 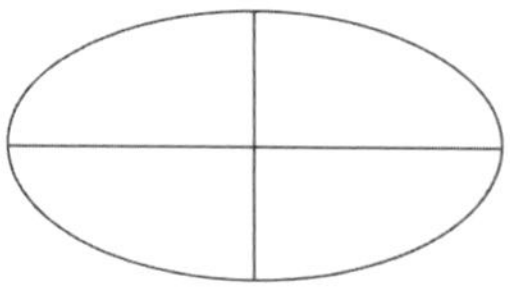   c

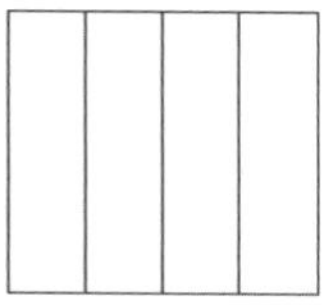

d 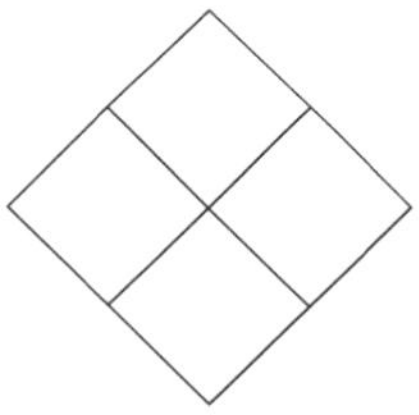   e 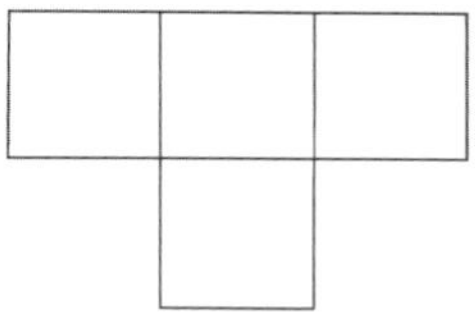   f

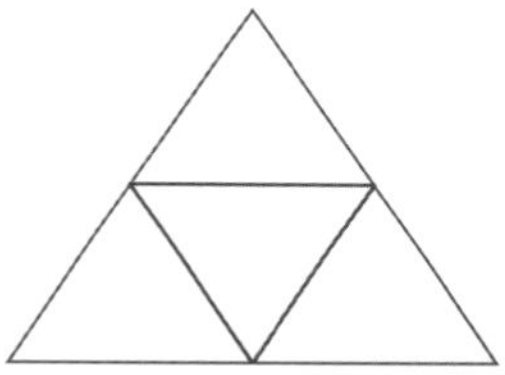

9 Draw lines to match the quarter to the whole shape.

quarter of a shape	whole shape

10 Fold a sheet of paper into quarters. Can you do this in more than one way? Try folding squares, circles, triangles and rectangles.

Problem-solving

11 Mrs Brown wants to make a patchwork quilt.
She cuts pieces of fabric like this to make squares.

a Is each piece a quarter or a half of a square? _____

b How many pieces does she need to cut to make 4 squares?

_____

c Mrs Brown decides to cut each triangle in half. How many triangles will she need to make up one square?

_____

Explain

Writing halves and quarters

When we divide things up into equal parts, we call each part a **fraction** of the whole.

 This is a whole square.

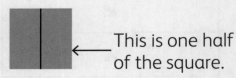 This is one half of the square.

There are 2 equal parts.

Each half is the same size. It is 1 of 2 equal parts.

We write one half as $\frac{1}{2}$. The line means **of**.

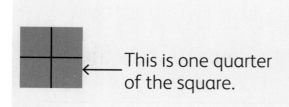

This is one quarter of the square.

There are 4 equal parts.

Each quarter is the same size. It is 1 of 4 equal parts.

We write one quarter as $\frac{1}{4}$.

12 What fraction of each shape is shaded? Write $\frac{1}{2}$ or $\frac{1}{4}$.

a

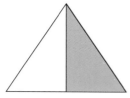

b

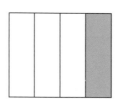

c

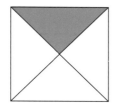

d

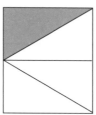

e

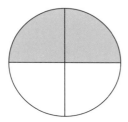

f

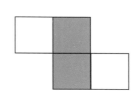

g

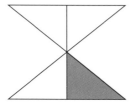

h

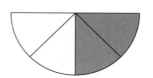

i

Investigate

13 Work in pairs. Use blocks of different colours to show these fractions. Explain what you did.

a $\frac{1}{2}$

b $\frac{1}{4}$

Problem-solving

14 Make up a story problem to match this number sentence.

$$\frac{1}{2} + \frac{1}{2} = 1$$

_____

_____

What did you learn?

Draw pictures to show these fractions. Label your pictures.

1 $\frac{1}{2}$

2 $\frac{1}{4}$

Topic 5 Review

Key ideas and concepts

Comparing sets and numbers

* When sets have the same number of objects, they are **equal**.
* = means **equals**
* > means **greater** or **more than**
* < means **less than**.

Ordinal numbers

* These are ordinal numbers: 1st, 2nd, 3rd, 4th, 5th, 6th, 7th, 8th, 9th, 10th.
* We use ordinal numbers to give the position of things.

Place value

* The **position** of a **digit** in a number tells us its **place value**. 17 = one ten and seven ones (10 + 7).

Fractions

* We can divide shapes and objects into two equal parts.
* Each part is called one half.
* Two halves make one whole.
* We write one half as $\frac{1}{2}$.
* We can divide shapes and objects into four equal parts.
* Each part is called one quarter or a fourth.
* Four quarters make one whole.
* We write one quarter as $\frac{1}{4}$.

Think, talk, write …

1   Write the following words in your mathematics dictionary. Explain what each word means in any way that you can.

   a half   b quarter

   c equal   d whole

2

Group the stars into tens and ones. Write the number in the table.

tens	ones

3  What fraction of this pie would you choose?

$\frac{1}{2}$ or $\frac{1}{4}$?

Give a reason.

Quick check

1 True or false? Write T or F.

a A B

Set A is = set B. ☐

b 7 > 6 ☐

c ☐

d The number 19 has one one and nine tens. ☐

e Each quarter of a square is the same size. ☐

f If you win the race, your position is 1st. ☐

Teaching notes for Topic 6: Algebraic thinking

Overview

In this topic students will learn to recognise and make different types of patterns using letters, shapes, colours, numbers and sounds.

A Different kinds of patterns	* Students need to recognise that patterns exist in many different ways. A pattern can be made with anything that is repeated. A sound or a colour can be repeated and so can shapes and numbers. They will learn that these can translate into other patterns.
B Number patterns	* Teaching students to recognise number patterns is important because it helps them to understand multiplication (and later, division). * Numbers make patterns that repeat and grow. Counting forwards or backwards can produce a pattern that grows. For example: 2, 4, 6, 8 … or 15, 10, 5, 0. * Students need to learn how to work out missing numbers in a pattern, and how to continue a number pattern. To do that they have to identify the pattern and work out whether the numbers are ascending or descending. * Skip counting produces a number pattern. We use skip counting to count groups of objects quickly. When you skip count, you miss out some numbers. You can skip count forwards and backwards and you can start counting at any point. * Learning to counting forwards in 2s (or 5s or 10s) is the same as repeatedly adding 2 (or 5 or 10). This will help students to do multiplication.

Mathematical vocabulary

Students need to know colours and number names in order to describe patterns. Revise these if necessary.

Notes for home

Your child is learning to recognise different types of patterns, including those made with sounds, shapes, colours, letters and numbers. There are many opportunities to find patterns around you and to ask your child to describe them. Counting in groups helps to develop a sense of number pattern – ask your child to count in 2s, 5s and 10s using body parts, number of wheels on cars, coins and any other objects in real life.

Algebraic thinking

A **Different kinds of patterns** *(pages 82–83)*

Describe what you see in the picture. Can you see the pattern? How was the pattern made?

Describe the pattern below.

Can you use the same shapes and colours to create another pattern?

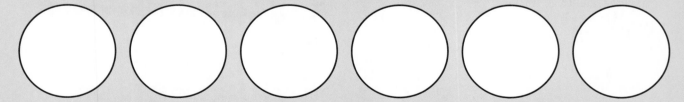

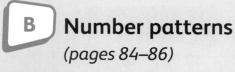

B **Number patterns**
(pages 84–86)

How many legs does a starfish have? How could you count the legs of four starfish quickly?

Taryn used a number line to count quickly to 20.

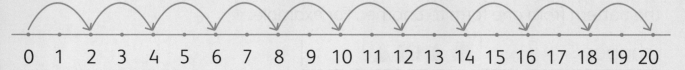

0 1 2 3 4 5 6 7 8 9 10 11 12 13 14 15 16 17 18 19 20

* Describe the number pattern you can see on the number line.
* Say the numbers in the pattern.
* What do you notice?

A Different kinds of patterns

Explain

We can make **patterns** in different ways.
We can use **colours**, actions, sounds, letters or numbers to make patterns.

Here are some different patterns.

The honeycomb in which bees make their honey has **shapes** that are repeated in a pattern.

These girls make a pattern of sounds when they clap their hands in different ways.

The tiles on the wall are arranged in a pattern. The same shapes are repeated.

Maths ideas

In this unit you will
* find different kinds of patterns
* complete patterns
* draw your own patterns.

Key words

patterns	shapes
colours	translate

We can represent the same pattern in different ways. We say we **translate** the pattern from one form to another. For example:

shapes	■ ▲ ■ ▲
colours	● ● ● ●
numbers	**3 4 3 4**
sounds	do re do re
letters	**A B A B**

These are all ways of showing the same pattern.

1 Draw lines to match the patterns.

2 Translate each pattern into sounds or actions.

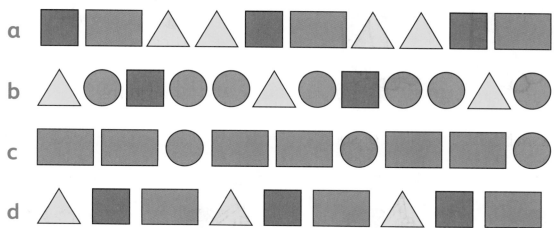

3 Translate these letter patterns into colour and shape patterns.

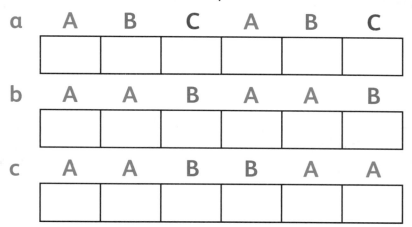

What did you learn?

Are these patterns the same?

A B B A B B

B Number patterns

When you **count** groups of things you **skip** some of the numbers so that you can count more quickly.

Maths ideas

In this unit you will
* use counting **patterns**
* count in 2s, 3s, 5s and 10s.

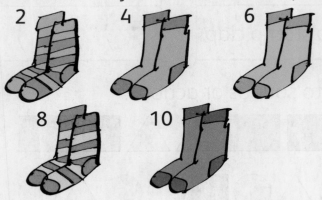

There are two socks in a pair. We can count the socks in 2s. The total number of socks is 10.

Tricycles have three wheels. We can count the wheels in 3s. The total number of wheels is 12.

A starfish has five legs. We can count the legs in 5s. The total number of legs is 25.

These crabs have ten legs each. We can count the legs in 10s. The total number of legs is 40.

Look at the number chart. Answer the questions.

1	2	3	4	5	6	7	8	9	10
11	12	13	14	15	16	17	18	19	20
21	22	23	24	25	26	27	28	29	30
31	32	33	34	35	36	37	38	39	40
41	42	43	44	45	46	47	48	49	50

1   Start with 2. Colour in every second number on the chart.

2   Start with 3. Circle every third number on the chart.

3   Count backwards in 2s from 10. Fill in the missing numbers.

   10 _____ 6 _____ _____

4   Count in 5s. Fill in the missing numbers.

   a 0   5 _____   15 _____   25

   b 15 _____   _____   30 _____

   c 35 _____   45 _____

   d 50   45   40 _____   30 _____

Problem-solving

5   The Level 1 class counted some objects in
   their class.

   a   Did they count in 3s or 5s? _____

   b   Can you fill in the totals?

Objects	Counted	Total
Pencils	ℋℋ ℋℋ ℋℋ ℋℋ	
Tables	ℋℋ	
Books	ℋℋ ℋℋ ℋℋ ℋℋ ℋℋ ℋℋ ℋℋ ℋℋ ℋℋ	

6 Circle the number that is incorrect in each number pattern.

    a 3, 6, 9, 10, 12

    b 2, 4, 6, 9, 10, 12

    c 10, 20, 30, 45, 50

    d 5, 10, 15, 21, 25

7 You are in a group of seven students. How could you count the numbers of toes and fingers in your group quickly?
Show how you counted.

_____

8 Tyrone was skip counting the shoes in the class. He reached a total of 45. Do you think his answer was correct? Why?

_____

What did you learn?

1 What is the rule for each of these counting patterns?

    a 10, 12, 14, 16, 18, 20

    Count in ⬚ s.

    b 35, 40, 45, 50

    Count in ⬚ s.

2 Choose a number from this counting pattern. Don't say the number, but explain it to your partner so that your partner can work out what the number is.
10   20   30   40   50

Topic 6 Review

Key ideas and concepts

* We can make patterns with colours, actions, sounds, letters or numbers.

* When you count groups of things you skip over some numbers so that you can count more quickly.

* We can count in **2s**, **3s**, **5s** and **10s**.

Think, talk, write ...

1   Tell your partner how to skip count in 5s. Use this number line to explain.

0 1 2 3 4 5 6 7 8 9 10 11 12 13 14 15 16 17 18 19 20 21 22 23 24 25

2   Work out a clapping or drumming pattern with a partner. How could you show this pattern in shapes or in numbers? Show it here.

Quick check

1   Count in 5s to join the dots. Colour in the picture.

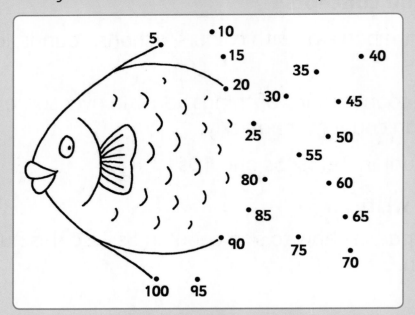

2   Finish each number pattern.

a | 12 | 14 | 16 | | | |

b | 50 | 40 | | | | |

c | 12 | 15 | 18 | | | |

d | 45 | 40 | 35 | | | |

3   Look around inside or outside your classroom.
    Find a pattern.
    Draw the pattern here and then describe it.

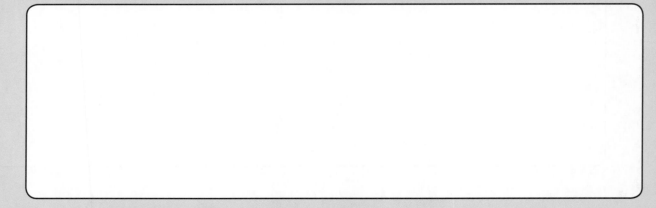

Teaching notes for Topic 7: Computation (2)

Overview

In this topic students will build on their knowledge of addition. They will learn to do repeated addition in preparation for learning how to multiply.

A Adding many times	* Students will need to understand that addition and multiplication are linked. Adding the same number over and over again is the same as multiplication. * Teaching students to add the same number over and over again is called repeated addition, for example: 2 + 2 + 2 + 2 = 8. Use real objects to teach this. * Teach the students to write number sentences to show repeated addition, for example: 3 + 3 + 3 = 9.
B Multiplying	* The students then need to be shown that 2 + 2 + 2 + 2 = 8 is the same as $2 \times 4 = 8$. Do this with real objects, showing the students how to group things and add groups together several times. * Teach the students how to write number sentences to show multiplication. Introduce the symbol we use to show multiplication ($\times$). For example: $3 \times 3 = 9$.

Mathematical vocabulary

Make sure that the students don't talk about multiplication 'sums' as this is incorrect. Rather call them calculations or multiplications. Students also need to learn to use the words **repeated addition**, **multiply** and **times**.

Notes for home

Your child is learning that addition and multiplication are related, for example, 3 + 3 + 3 = 9 and 3 times 3 = 9. The idea of multiplication is being taught through skip counting and by using objects to make an array with rows and columns.

You can use dried beans or other small objects to help your child understand this. Make an array with two rows of four beans and ask, 'What is two times four?' Or ask your child to model a multiplication with the beans (using small numbers and totals up to 12 only). Say things like, 'Show me how to work out 3 times 4 using the beans.'

1234

A **Adding many times** *(pages 92–94)*

How many windows can you see? How can you count them quickly without counting each one?

| 0 | 1 | 2 | 3 | 4 | 5 | 6 | 7 | 8 | 9 | 10 | 11 | 12 |

Show how you can use this number line to add the same number over and over again to find the answer.

2 + 2 + 2 + 2 + 2 + 2 = ☐

What did you do? Did you skip count in 2s or 3s? ☐

1234

B **Multiplying** *(pages 95–98)*

How many sets of hearts are there? How many hearts in each set?

A square has 4 sides. How many sides are there on 3 squares?
Tick all the number sentences that describe this.

$4 + 4 + 4 = 12$ ☐

Three 4s are 12. ☐

Three sets of four equals 12. ☐

$3 \times 4 = 12$ ☐

What do you notice? ☐

☐

A Adding many times

Look at the stickers.

There are 2 stickers in each row.

There are 4 rows.

We can add 2 each time to find out how many there are.

$2 + 2 + 2 + 2 = 8$.

We call this **repeated** addition.

This is the same as skip counting in 2s.

Maths ideas

In this unit you will
* **add** the same numbers over and over again
* double numbers
* write addition **number sentences**.

Key words

add

number sentences

repeated

number line

1 Copy and complete the number patterns.

a Add 2 each time.

2 ▷+2▷ 4 ▷+2▷ ___ ▷+2▷ ___ ▷+2▷ ___ ▷+2▷ ___ ▷+2▷ 14

b Add 3 each time.

3 ▷+3▷ 6 ▷+3▷ ___ ▷+3▷ ___ ▷+3▷ ___ ▷+3▷ ___ ▷+3▷ 21

2 Count in groups. Fill in the number sentences.

    a Count in 3s.

    _____ 3s are _____

    b Count in 4s.

    _____ 4s are _____

    c Count in 5s.

    _____ 5s are _____

    d Count in 10s.

    _____ 10s are _____

3 Count in 5s from 0. Show the hops on the **number line**.

```
0  1  2  3  4  5  6  7  8  9  10 11 12 13 14 15 16 17 18 19 20
```

4 Complete the number sentences.

    a 3 sets of 5 = 15

       5 + 5 + 5 = 15

    b ____ sets of ____ = ____

       ____ + ____ + ____ + ____ = ____

93

c _____ sets of _____ = _____

_____ + _____ + _____ = _____

d _____ sets of _____ = _____

_____ + _____ + _____ + _____ = _____

5 Fill in the missing numbers.

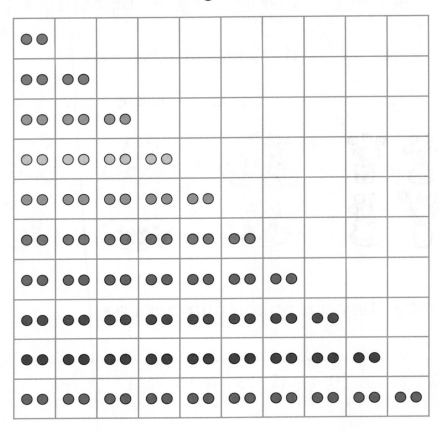

1 set of 2 = 2

2 sets of 2 = 4

_____ sets of 2 = _____

_____ sets of 2 = _____

_____ sets of 2 = _____

_____ sets of 2 = _____

_____ sets of 2 = _____

_____ sets of 2 = _____

_____ sets of 2 = _____

_____ sets of 2 = _____

What did you learn?

Fill in the missing numbers.

☐ sets of ☐ = ☐

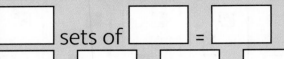

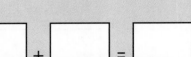

B Multiplying

Explain

When you add 2 + 2 + 2 + 2 = 8, you are adding the same number (2) four times.
Think of this as 'four times two'.

We use the symbol × when we **multiply**.
4 × 2 = 8   means 4 **times** 2 equals 8.

Maths ideas

In this unit you will
* write **multiplication** number sentences
* solve and write multiplication problems.

Important symbols

× multiply

1   Complete the sentences to describe the sets.

a

_____ sets.

_____ in each set.

_____ + _____ = _____

2 × _____ = _____

b

_____ sets.

_____ in each set.

_____ + _____ = _____

2 × _____ = _____

c

_____ sets.

_____ in each set.

_____ + _____ = _____

2 × _____ = _____

d

_____ sets.

_____ in each set.

_____ + _____ = _____

2 × _____ = _____

Key words

multiplication
multiply
times

e

_____ sets.

_____ in each set.

_____ + _____ = _____

2 × _____ = _____

f

_____ sets.

_____ in each set.

_____ + _____ = _____

2 × _____ = _____

2 Complete the sentences to describe the sets.

a

_____ sets.

_____ in each set.

_____ + _____ + _____ = _____

3 × _____ = _____

b

_____ sets.

_____ in each set.

_____ + _____ + _____ = _____

3 × _____ = _____

c

_____ sets.

_____ in each set.

_____ + _____ + _____ = _____

3 × _____ = _____

d

_____ sets.

_____ in each set.

_____ + _____ + _____ = _____

3 × _____ = _____

3 Complete the sentences to describe the sets.

a

_____ sets.

_____ in each set.

___ + ___ + ___ + ___ = ___

4 × _____ = _____

b

_____ sets.

_____ in each set.

___ + ___ + ___ + ___ = ___

4 × _____ = _____

c

_____ sets.

_____ in each set.

___ + ___ + ___ + ___ = ___

4 × _____ = _____

d

_____ sets.

_____ in each set.

___ + ___ + ___ + ___ = ___

4 × _____ = _____

4   Make up a multiplication sentence for each set.

a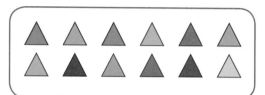

_____ × _____ = _____

b

_____ × _____ = _____

c

_____ × _____ = _____

d

_____ × _____ = _____

Problem-solving

5   Tess plants seedlings in the garden. She plants
    two rows with six seedlings in each row.
    How many seedlings does she plant?

    _____ × _____ = _____

6   Peter blows up five balloons. Then he blows up five more
    balloons. How many balloons does he blow up?

    _____ × _____ = _____

7 Write two problems for your partner to solve.
Use these multiplication sentences.

a 3 × 4 = ?

_____

_____

b 2 × 4 = ?

_____

_____

Investigate

8 How can you use a calculator to get this answer? 6 × 1 = ☐

9 What do you do when you double a number? Try doubling
these numbers. Write the number sentences.

a double 2:   2 + _____ = _____

b double 3:   3 + _____ = _____

c double 5:   5 + _____ = _____

What did you learn?

Draw pictures to show each number sentence.

1 3 + 3 + 3 = ☐

2 10 + 10 + 10 = ☐

3 3 × 4 = ☐

Topic 7 Review

Key ideas and concepts

✳ **Repeated** addition is the same as skip counting.

✳ **Multiplication** is a short way of doing repeated addition.

✳ We use the symbol × to show that one number is multiplied by another.

Think, talk, write ...

1   Adding 3 over and over again is called ☐ addition.

3 + 3 + 3 + 3 = ☐
We can also write this as 3 ☐ 4 = ☐ .

2   Write the word multiplication in your mathematics dictionary. Write a definition of the word.

Quick check

1   Martine picked three flowers for her mother, three flowers for her sister and three flowers for her friend.

   a   How many flowers did Martine pick? Draw a picture to show this.

   b   Complete the number sentence. ☐ × ☐ = ☐

2   You have a bag with 12 beads. You want to count the beads quickly. Arrange the beads in rows in two different ways to show how you can do this. Draw pictures and write a number sentence to go with each picture.

Teaching notes for Topic 8: Measurement (1)

Overview

In this topic students will first build on their knowledge of measurement in various ways; they will compare lengths and heights using words before they estimate measurements using non-standard units. After that they will learn about accurate measurement, using the metre as a unit of measurement. They will also learn to measure distances.

A Estimate and compare lengths and heights	✳ Length is a measurement of how long something is from end to end. Distance, height, width (breadth), thickness and depth are all measures of length. ✳ You could start by letting students make comparisons of the lengths of real objects and heights of people. Let the students use words such as short, tall, long, shorter than and longer than to make these comparisons.
B Use non-standard units to measure	✳ We can use non-standard units to measure lengths. Non-standard units are not accurate, but they help us to estimate measurement. You can teach the students to use armspans or handspans. ✳ Armspans: A child's armspan is from the fingertips of one outstretched arm to the fingertips of the other outstretched arm. ✳ Handspans: A handspan is from the tip of the thumb to the tip of the little finger on an outstretched hand. ✳ You can also teach the students to use items such as paperclips and bottle tops to measure. Make sure that the items are the same size though, and that there are no gaps between the items. ✳ It is important to teach students to estimate measurements. Estimating helps us to do quick measurements and to check real measurements. It takes practice to develop this skill. Use real objects or pictures.
C Estimate and measure length and height in metres	✳ The metre is the base unit of length in the metric system. ✳ Teach the students to take accurate measurements using metre sticks, measuring tapes and rulers.
D Estimate and measure distances	✳ Teach the students how far away things are from each other (distance) using objects and places in the classroom and around the school.

Mathematical vocabulary

Students need to learn to use vocabulary such as **metre**, **longer than**, **shorter than**, **height**, **length** and **distance**. They also need to understand the concept of **estimating**.

Notes for home

Your child is learning how to estimate and take measurements in metres.

Encourage your child to take simple measurements at home. Measure a height of 1 m on a wall and then let members of the family stand in front of the measurement. Let your child say who is taller or shorter than 1 m.

 Estimate and compare lengths and heights *(pages 104–106)*

How long is the lizard's tail? Is its tail longer or shorter than its body? How could you find out?

* Which is shorter, a pencil or a cricket bat?
* What is the shortest thing in your classroom?
* Are you shorter or taller than your teacher?
* What is the tallest building in your community?

 Use non-standard units to measure *(pages 107–109)*

How could you use these things to measure? What else could you use?

* Which parts of your body can you use to measure things?
* How can you describe your measurements?

C Estimate and measure length and height in metres (pages 110–111)

> What are these? What do we use them for? What do the numbers tell us?

* Use your fingers to measure the length of your pencil. How many fingers did you use?
* Now use your ruler to measure your pencil. Read the number on the ruler. What is the measurement on the ruler?

D Estimate and measure distances (pages 112–113)

> Which desk is nearer to the door? Which desk is nearer to the window? How could you measure this?

* How far is your classroom from other buildings at the school?
* How could you measure this distance?

A Estimate and compare lengths and heights

Your **height** is how tall you are.
The **length** of a ladder is how long it is.
We can put things next to each other to compare their length and height.

The yellow ribbon is **short**.
The red ribbon is long.
The orange ribbon is **longer than** the red ribbon.

Anna is **tall**. Her brother is short.
Anna is **taller than** her brother.

1   Underline the **shortest** child. Circle the **tallest** child.

2   Draw a tree that is taller than this tree.

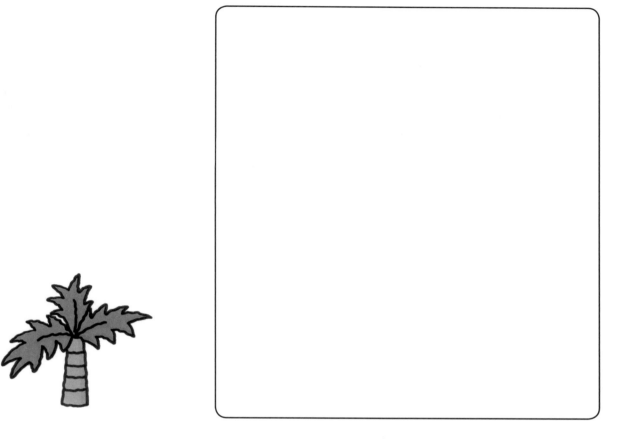

3   Colour the longest object red. Colour the shortest object blue.

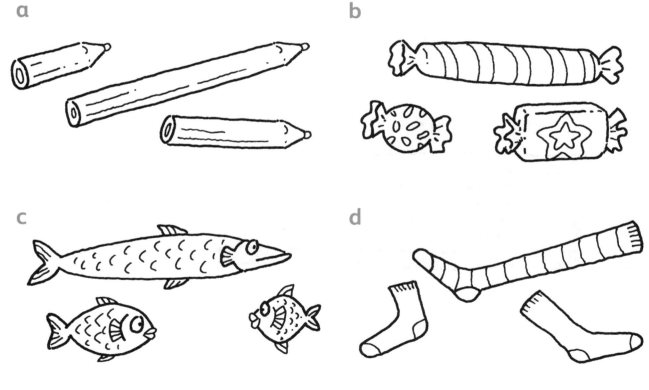

4   Colour the tallest green. Colour the shortest blue.

a

b

c

d

Investigate

5   What objects are longer than your thumb? Use your thumb to measure.

6   What objects are the same length as your foot? Choose an object. Use your foot to measure it. Describe the object.

What did you learn?

A

B

C

D

Look at the ribbons. Arrange the ribbons in order from the shortest to the longest. Write the letters in the box below.

B Use non-standard units to measure

Explain

We can use objects to **measure** the **length** or **height** of other objects.

We measured the length of this book using paperclips. We put the paperclips end to end with no gaps in between and no overlaps.

Maths ideas

In this unit you will
* use objects and parts of your body to measure objects
* **estimate** the length of objects.

Key words

measure

length

height

estimate

1   Use paperclips or other objects to measure the length of each feather.

a

_____ paperclips

b

_____ paperclips

2 Draw lines.

a Draw a line that is longer than two paperclips.

b Draw a line that is the same length as three paperclips.

Explain

We can use parts of our body to **measure** length and height.

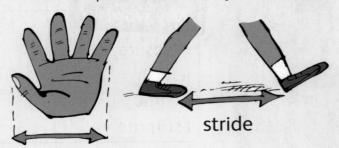

handspan   stride

armspan

It is useful to be able to **estimate** the length or height of an object. This means we guess the length, based on what we know.

If we know more or less how wide our handspan is, we can estimate the length of an object.

This table is about 3 handspans long.

---

3 Use your fingers to measure these ribbons. First estimate how long the ribbons are. Then use your fingers to measure.

a

Estimate: _____ fingers

Measurement: _____ fingers

b

Estimate: _____ fingers

Measurement: _____ fingers

Investigate

4   Work in pairs. Which body measurement could you use to measure these lengths? Why?

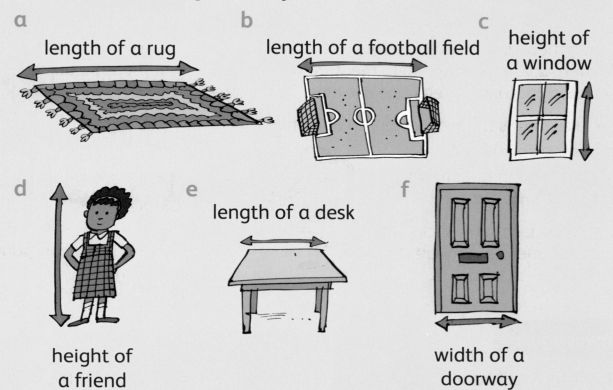

a   length of a rug

b   length of a football field

c   height of a window

d   height of a friend

e   length of a desk

f   width of a doorway

5   Jo and her big brother needed to measure the length of the floor because they wanted to get a new rug. They used strides to estimate the measurement, but they got different answers. Why?

_____

What did you learn?

Measure this pencil. Use paperclips.

Estimate: ☐ paperclips   Measurement: ☐ paperclips

C Estimate and measure length and height in metres

Explain

A **metre** is a standard **unit** of length. This means that a metre is always the same. It is more accurate to **measure** in metres than to measure with parts of your body because different people are different sizes!

We use metres to measure length and height.

m is short for metre.

A desk is about one metre long. We write 1 m.

We can use a metre stick to measure metres. A metre stick is exactly 1 m long.

Maths ideas

In this unit you will
* **estimate** and measure objects using metres
* **compare** measurements.

Key words

metre	estimate
unit	compare
measure	

Investigate

1 Make a measuring tool that you can use to measure metres. Here are some ideas:
   * rolled-up sheets of newspaper
   * piece of rope or string.

How could you check that your measuring tool measures exactly 1 m?

2 Work in pairs. Find these objects in and around your classroom. Estimate the length or height in metres. Then use a metre stick to measure them accurately.

a

Estimate: ____ m
Measurement: ____ m

b

Estimate: ____ m
Measurement: ____ m

c

Estimate: ____ m

Measurement: ____ m

d

Estimate: ____ m

Measurement: ____ m

3   Find three things in your classroom that are:

a   shorter than 1 m

_____   _____   _____

b   longer than 1 m

_____   _____   _____

c   about 1 m long

_____   _____   _____

Problem-solving

4   Nadia and Ben used metre sticks to measure the height of a door in their classroom.
Nadia measured 2 m.
Ben measured more than 2 m.
What happened? Why did they get different answers?

_____

What did you learn?

1   What is the tallest thing in your school?

2   Estimate how tall it is in metres.

D Estimate and measure distances

Explain

The **distance** between two objects is how far away they are from each other. We can **measure** distances in **metres**.

Maths ideas

In this unit you will
* **estimate** and measure distances around your school
* compare measurements.

1   Work in pairs. Estimate and then measure these distances at your school.

Key words

distance   metres

measure   estimate

   a   The distance from your desk to the door.

Estimate: _____ m

Measurement: _____ m

   b   The distance from your classroom to the Level 2 classroom.

Estimate: _____ m   Measurement: _____ m

   c   The distance from your classroom to the playground.

Estimate: _____ m   Measurement: _____ m

   d   Put the distances in order from the shortest to the longest.

_____ m   _____ m   _____ m

2   Look at the pictures. Which distance is longer? Underline the distance that is longer.

a

A   from the classroom to the toilets

B   from the classroom to the office

b

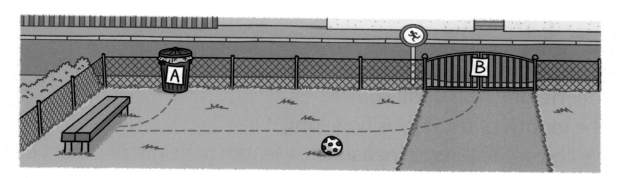

A from the bench to the bin   B from the bench to the gate

4   Think about the distance around a tennis
    court or basketball court.

   a   Which body measurement would you
        use to help you estimate the distance? _____

   b   Why would you use this unit of measurement?

     _____

   c   What could you use to make an accurate measurement?

     _____

Challenge

5   What units do we use to measure longer distances,
    kilometres or kilolitres?

     _____

What did you learn?

Complete the sentences.

1   We can measure the [   ] from one place

    to another in [   ].

2   We can [   ] distances by using strides.

Topic 8 Review

Key ideas and concepts

* Your **height** is how **tall** you are.

* The **length** of a ladder is how **long** it is.

* We can use objects to **measure** the length or height of other objects.

* We can use body measurements (handspan, stride, armspan) to estimate measurements.

* When we **estimate** the length or height of an object, we guess what that measurement may be based on what we know.

* We use **metres** to measure length, height and distances.

* **m** is short for metre.

* The **distance** between two objects is how far they are from each other.

Think, talk, write ...

1   What words could match these definitions? Write the words and the definitions in your mathematics dictionary.

   a   a unit that we use to measure length, height and distance

   b   to use what information you know to guess the measurement of something

2   Think of three ways in which you can measure things in your classroom without using a ruler or a metre stick. Share your ideas with your group.

3   Work in pairs. Give your partner a tip about how to measure something.

Quick check

1   Measure this ribbon. Use paperclips.

   Estimate: _____ paperclips   Measurement: _____ paperclips

2   Which is taller? Underline the answer.

   the door in your classroom   the wall in your classroom

Teaching notes for Topic 9: Data handling (1)

Overview

In this topic students will learn how to read and interpret simple pictographs. Then they will start to learn how to collect data in order to make their own pictographs.

A Picture graphs	
	* Students need to understand that data is information that we have collected.
	* Picture graphs (pictographs) use small images or symbols to represent data. Each picture stands for a number on the graph. For example, ☺ could represent one person. A 🚌 could represent one person who comes to school by bus every day. A pictograph has a key that explains what the pictures or symbols on the graph represent.
	* Students need to learn that in order to understand a graph, they need to look for clues that tell them what the data on the graph is about. Help them to read the title of a pictograph and to look at the key before they answer questions about the data on the graph.

B Collecting data and making pictographs	
	* Help students to understand that we collect data in order to answer questions about our environment and to make choices. For example, if we read a graph about weather patterns, we can judge what clothes to take when we visit a different place.
	* Teach the students to start collecting simple data by observing things or by asking people questions.
	* Help the students to record data accurately. The tally system is accurate and easy to use when students understand how it works.
	* Teach the students how to display the data they have collected on simple pictographs.

Mathematical vocabulary

Students need to be able to understand the following words: **picture graph (pictograph)**, **key**, **data**, **information**, **tally** and **symbol**.

Notes for home

Your child is learning to sort objects and make tally charts and graphs to show how many are in each group.

You can help your child understand data and graphs by asking them to sort items into groups based on the way in which they are alike – sort socks by length or by colour, sort cutlery and crockery by type. You can also play 'tally' games when you are driving. Choose three car colours or models of cars, for example, and let your child use tallies to record how many of each they see on the journey.

 Picture graphs *(pages 118–121)*

Can you name the fruit in the bowl? How many of each fruit?

Fruit in a fruit bowl	Number of fruits

* How can you show the information about the fruit bowl on this chart?

* Complete the chart.

* What kind of chart is this?

What information did the students collect?
How did they get the information?

* Why is this information useful?
* How could the children show this information to other people?

A Picture graphs

Explain

A **pictograph** is a **picture graph** that shows information we have collected. We call this information **data**.

We group information into **sets of data** to make pictographs.

Maths ideas

In this unit you will
* talk about data
* read pictographs.

Key words

pictograph	sets
picture graph	title
	symbol
data	key

How to read the information in a pictograph

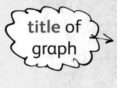

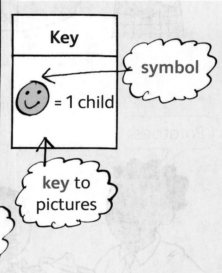

1 Look at the pictograph above and answer the questions.

a How many students were absent on Monday?

b How do you know this? _____

c On what day were three students absent?

d On what day was only one student absent?

2 Look at this pictograph and then answer the questions.

   a What does this pictograph show?

Favourite fruits

_____

   b What does  represent?

_____

   c How many children like mangoes best?

_____

   d Which fruit did most children choose?

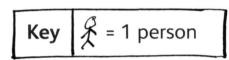

_____

3 Look at this pictograph and then answer the questions.

_____

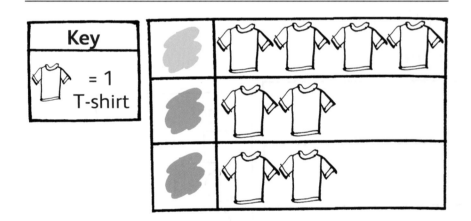

   a Write a title for the graph.

   b How many yellow T-shirts are there? _____

   c What can you say about the green T-shirts and the pink T-shirts?

_____

   d How can you collect data like this?

_____

4 Look at the shapes in the box.

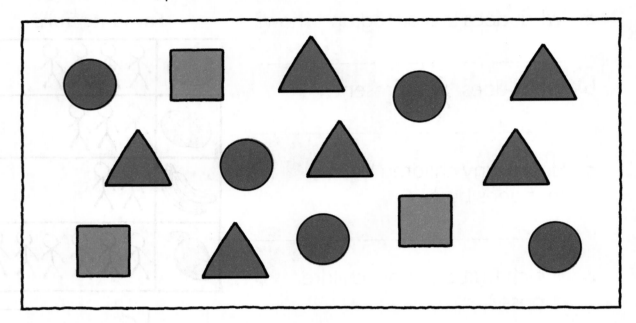

a How many shapes are there in the box?

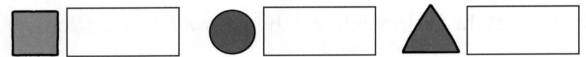

b Complete this pictograph.

Number of shapes

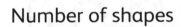

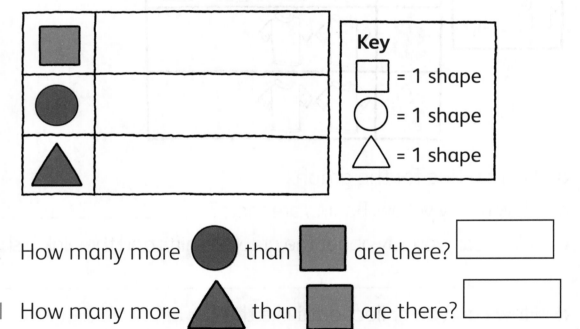

5 Look at the picture of fruits at the market. Complete the pictograph.

apples		pears		**Key**
plums		bananas		● = 1 fruit
mangoes		avocados		

What did you learn?

Answer the questions about this pictograph.

How my friends go to school

car	🏃 🏃
bus	🏃
walk	🏃 🏃 🏃 🏃

Key

🏃 = 1 friend

1 🏃 🏃 = ☐ friends

2 How many friends walk to school? ☐

3 What type of transport does only one friend use? ☐

B Collecting data and making pictographs

Explain

Collecting data

When you collect **data** the first thing you do is to decide what information you want to find out. Then you decide how you will find this out. Here are two ways you can collect data.

1 You can ask questions to collect data.
   * Think of the question you need to ask. *What is your favourite fruit? How old are you? Where do you live? How do you get to school?*
   * Ask different people the same question.
   * Record the answers. For example:

   Favourite fruits

apples	✓ ✓ ✓ ✓
bananas	✓ ✓ ✓ ✓ ✓
lemons	

2 You can observe and make notes to collect data.

   Imagine you want to find out how many buses stop at the bus stop on Saturday mornings. To do this you can stand and count them. You can record each bus as you see it. **Tally** marks are useful for keeping count.

   * A tally is a mark that you make to keep count. / means 1. // means 2. When you reach 5, you cross through the four tallies to make a group of five, like this: *####*. You can count in fives to get a total.
   * Fill in the number shown by each set of tally marks.

   /// ☐   //// ☐   #### ☐   #### // ☐

   * Draw tallies to show each amount.

   ☐ 8   ☐ 6   ☐ 10

   Number of buses on Saturdays

buses	#### ///

Maths ideas

In this unit you will
* collect data
* use pictures to show data on a graph.

Key words

data	key
tally	pictograph
symbol	

1 Find out which drink your friends like.

Ask 15 friends.

Each person should give only one answer.

Use tallies to record the answers.

What do you like to drink?

water	
milk	
fruit juice	
tea	

2 Complete the pictograph to show your data.

Title: _____

water	
milk	
fruit juice	
tea	

Pictograph tips

Choose a **symbol** to show one person.

Draw the symbol in the **key**.

Give your **pictograph** a title.

Key

☐ = 1 person

3 Look at the picture of the animals on the farm. Complete the pictograph to show this information.

a Choose a symbol to use in the key. Fill in the key.

b Count the groups of animals. Draw symbols to show how many there are.

c Give the graph a title.

Title: _____

dogs	
ducks	
horses	
chickens	

Key

= 1 animal

4 **a** Count the letters in each name.

Write the number in the circle.

Jamila ◯	Tyrone ◯	Janae ◯
Max ◯	Jovan ◯	Curtis ◯
Celine ◯	Nicholas ◯	Kimberly ◯
Rayshawn ◯	Serenity ◯	Clara ◯

b Complete the pictograph for the names in Question 4a.

Letters in names

less than 6	
6	
more than 6	

Key

🚶 = 1 child

c Write your name. _____

d Add your own data to the graph.

5   Mel, Bob and Jo are friends.

   ✳ Mel drinks 5 glasses of water every day.

   ✳ Jo drinks 7 glasses of water.

   ✳ Bob drinks one glass more than Jo.

   ✳ Which pictograph shows this information? Tick it.

Mel	🥛 🥛 🥛 🥛 🥛
Bob	🥛 🥛
Jo	🥛 🥛 🥛 🥛 🥛 🥛 🥛

☐

Mel	🥛 🥛 🥛 🥛 🥛
Jo	🥛 🥛 🥛 🥛 🥛 🥛 🥛
Bob	🥛 🥛 🥛 🥛 🥛 🥛 🥛 🥛

☐

6   Read the information.

   There is a sports club in town. These are the sports that people can do at the club:

   ✳ swimming

   ✳ cricket

   ✳ running

   ✳ basketball

   ✳ Last week 15 people went swimming.

   ✳ 22 people played cricket.

   ✳ 16 people went running.

   ✳ No one played basketball.

   Show this information on the pictograph.

People at the club last week

Problem-solving

swimming	

Key

= 1 person

What did you learn?

Use the following words to complete the sentences.

key	title	data	pictograph

1 We collected [_____] to find out how much water people drank each day.

2 We drew a [_____] to show this information.

3 The [_____] has symbols that show what the pictures represent.

4 The [_____] tells us what the graph is about.

Topic 9 Review

Key ideas and concepts

* A **pictograph** is a picture graph that shows information that we have collected. We call this information **data**.

* A pictograph must have a **title**.

* The **key** tells you what each picture means.

* We can use **tallies** to keep count.

* A tally is a small mark like this /. Every 5th tally is drawn across the others to make a group of 5 like this ////.

Think, talk, write …

1   Write three new mathematics words in your mathematics dictionary. Draw a picture to show what each word means.

2   Work in groups. Discuss how you can collect the following information and how you can show it on a graph.
   You want to make a class flag. You need to find out which colours people in the class like best.

Quick check

1   Answer the questions about Jo's pictograph.

   The shells I collected at the weekend

Saturday	🐚 🐚 🐚 🐚 🐚 🐚 🐚 🐚 🐚
Sunday	🐚 🐚 🐚 🐚 🐚 🐚 🐚 🐚 🐚 🐚 🐚 🐚 🐚 🐚 🐚 🐚 🐚 🐚

   Key

   🐚 = 1 shell

   a   How many shells did Jo collect on Saturday?

   b   On which day did she collect more shells?

   c   How many shells did she collect altogether?

Test yourself (2)

1 Write these numbers in the correct boxes: 4th, 1st, 3rd, 6th, 5th, 2nd.

2 Tick the correct box.

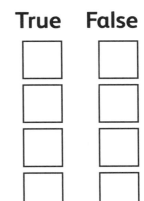

	True	False
a The girl in yellow was 1st.		
b The girl in purple was 2nd.		
c The girl in white was 6th.		
d The girl in red was 5th.		

3

Group the mangoes into 10s and 1s.
Write the number in the table.

tens	ones

4 Colour the shapes to show each fraction.

a
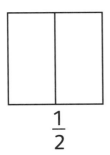
$\frac{1}{2}$

b
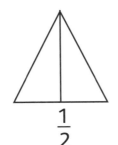
$\frac{1}{2}$

c
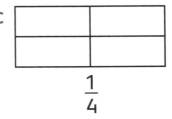
$\frac{1}{4}$

d
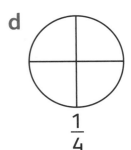
$\frac{1}{4}$

5 Join the dots. Count in 2s. Start at 0.

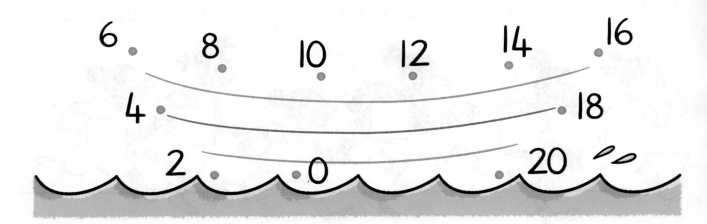

6 Complete the number sentences.

    a $2 + 2 + 2 =$ _____

    b $2 \times 3 =$ _____

    c $5 + 5 + 5 +$ _____ $= 20$

    d $5 \times$ _____ $= 20$

7 How long is a table?

    a Estimate the length of a table in your classroom. _____

    b Then measure the table. Write the measurement. _____

8 Read the graph. Answer the questions.

Amount of water we drink

Delia	🥛🥛🥛🥛🥛
Tyrone	🥛🥛🥛
Jo	🥛🥛🥛🥛🥛🥛🥛🥛

Key

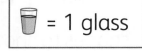 = 1 glass

    a How many glasses of water does Delia drink every day? _____

    b Who drinks the most water every day? _____

    c Does Tyrone drink more or less water than Delia? _____

Teaching notes for Topic 10: Measurement (2)

Overview

In this topic students will build on their knowledge of measuring time and money.

A Days, weeks and months	* Make sure the students are aware of the different times of the day (such as morning and evening) and that they can associate these times with daily activities. * Then teach the students to measure time in hours, days, weeks and months. They should learn to read and write the names of the days and the months.
B Telling the time	* We use analogue and digital clocks to tell time. Teach the students how to read analogue clocks first. Many students will be aware of digital clocks and watches, so include them in your inverstigations and discussions in class. * Analogue clocks: These are 12-hour clocks or watches. They have hands that show the time in hours and minutes. Some also show seconds, but you do not need to teach the students about seconds yet. * Digital clocks: These are 24-hour clocks that show the time in digits.
C Money	* We use notes and coins as money. This is called currency. Each country or group of countries has its own currency. * The currency in the Eastern Caribbean is dollars and cents. * Notes and coins are available in different denominations, for example: $1, 5¢, 10¢, 25¢ and 50¢. Students need to learn the values of these notes and coins.

Mathematical vocabulary

Students will need to learn words to talk about times of day (**morning** and **midday**, for example), the names of the days of the week and months of the year, words that give accurate time measurements (**hours**, **minutes**), words that describe analogue clocks (**hands**, **face**) and tell the time (**o'clock**). Students also need to be able to describe the value of some coins and notes.

Notes for home

Your child is learning about time at school. Being able to understand and describe time will help your child to understand and plan their everyday lives.

Help your child to tell the time at home. Use an analogue (12-hour) clock if you have one because this is what your child will learn first at school. Focus on hours and half hours at this stage. Make your child aware of the time and at what times things happen at home. Say things such as, 'We have supper at 7 o'clock. Is it supper time?', 'We are going to the beach in the afternoon.'

10 Measurement (2)

A **Days, weeks and months** *(pages 134–140)*

What time of day do you think this is? How can you tell? Do you enjoy this time of day? Why or why not?

1 Answer the questions in your groups.

* What is your favourite time of day?
  morning   afternoon   evening

* Which is longer – a day or a week?
  day   week

* Which is shorter – a month or a year?
  month   year

* When is your birthday?

* What is the date today?

B

This clock has hands that show us the time. What is the time? What do you do at this time?

Telling the time *(pages 141–144)*

Answer the questions by circling the correct time. Then share your answers with your group.

1. What time does Jo leave for school in the morning?

   8 o'clock   4 o'clock

2. What time does Jo have lunch?

   12 o'clock   6 o'clock

C   **Money** *(pages 145–147)*

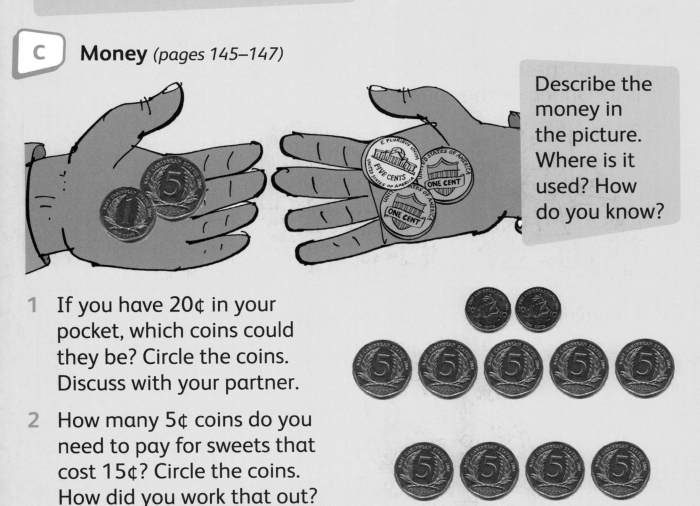

Describe the money in the picture. Where is it used? How do you know?

1. If you have 20¢ in your pocket, which coins could they be? Circle the coins. Discuss with your partner.

2. How many 5¢ coins do you need to pay for sweets that cost 15¢? Circle the coins. How did you work that out?

A Days, weeks and months

Time of day

Our days begin when the sun rises in the **morning**.

The middle of the day is called **midday**.

After midday it is **afternoon**.

When the sun goes down again, it is the **evening**.

Night follows, and it is dark.

Midnight is the middle of the night.

morning
↓
midday
↓
afternoon
↓
evening
↓
night
↓
midnight

In this unit you will
* talk about different times
* name the days of the week
* name the months of the year
* give and write the date.

morning
midday
afternoon
evening
night
midnight
day
week
schedule
month
year
calendar

1   Look at these pictures. What time of **day** is it? Underline the correct word.

a

morning   night

b

morning   night

c

afternoon   evening

2 Look at the pictures. Write a word next to each picture to show the time of day.

| morning | night | midday | afternoon |

a

_____

b

_____

c

_____

d

_____

3 Draw what you did at these times yesterday.

in the morning	at midday
in the afternoon	in the evening

Days and weeks

There are seven days in one **week**.

The days are:
* Sunday   * Monday   * Tuesday   * Wednesday
* Thursday   * Friday   * Saturday

4   Write the names of the days.

a   The day before Monday. _____

b   The day after Thursday. _____

c   The 2nd day of the week. _____

d   The 3rd day of the week. _____

e   The last day of the week. _____

f   The days on which you go to school. _____

_____   _____

_____   _____

5   Read Denzel's **schedule** with your teacher. Write your answers.

Sunday	Church. Lunch at Granny's.
Monday	Football practice.
Tuesday	Library.
Wednesday	Class trip to fire station.
Thursday	Choir practice. Football practice.
Friday	Fishing (with Dad!)
Saturday	Beach. Watch a movie.

a What did Denzel do on Monday?

_____

b On which day did he watch a movie?

_____

c How many days a week does he have football practice?

_____

d When did the class go to the fire station?

_____

6 Draw up your schedule for 3 days of the week.
   Fill in the names of the days.
   Write or draw what you will do each day.

My 3-day schedule

Challenge

7 Solve the riddles. Guess the days.
   a I am the only day of the week with 9 letters in my name.

   _____

   b I am before Thursday and after
      Monday, but I am not Wednesday! _____

Explain

Months

There are 12 **months** in one **year**.

The names of the months are:

January   February   March   April   May   June

July   August   September   October   November   December

8   Look at the **calendar**. Complete the sentences.

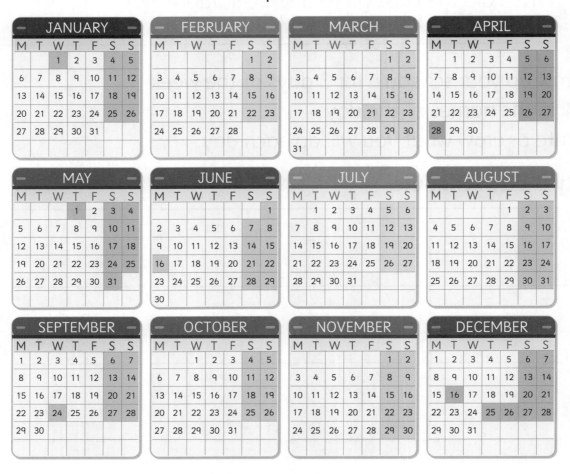

a   November has _____ days.

b   July has _____ days.

c   The month after January is _____.

d   The 10th month is _____.

e   The month before June is _____.

9 The months and days are mixed up. Colour the months blue and the days yellow.

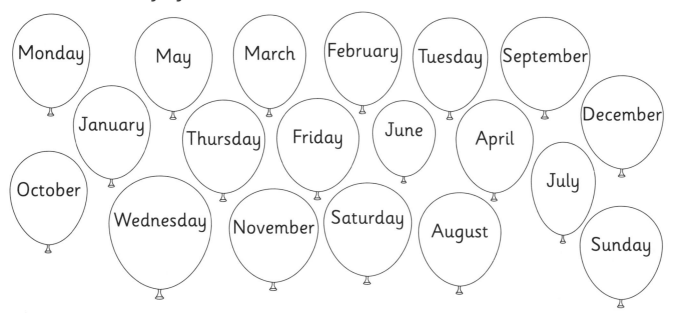

10 Here is a calendar for August.

August						
Sunday	Monday	Tuesday	Wednesday	Thursday	Friday	Saturday
						1
2	3					8
9	10	11	12			
16				20	21	22
	24	25				
	31					

a Fill in the dates that are missing.

b How many days are there in August? _____

c How many full weeks are there in August? _____

d On which day of the week did August start? _____

e What day of the week is the 3rd of August? _____

f Colour in all the Saturdays in August. How many are there?

_____

g   Some children have their birthday in August. Write the day that their birthday will fall on.

Daniel – 5th August _____

Tori – 17th August _____

Maika – 22nd August _____

h   Janae's birthday is on the third Thursday of August.

What date is her birthday? _____

_____

Investigate

11 How many students in your class have birthdays this month?

12 How can you find out?

13 How can you record this?

What did you learn?

1   Answer these questions.

a   How many days in a weekend?

b   How many days are left this week?

c   What day will it be tomorrow?

d   What day was it yesterday?

e   What is the month after April?

f   How many months in a year?

2   Make up your own quiz in groups. Swap questions with another group. Answer their questions.

B Telling the time

Explain

A clock shows the time in **hours** and **minutes**.
The short 'hand' on the **clock** points to the hour.
It is the hour hand.
The long 'hand' on the clock shows the minutes.
It is the minute hand.

The long hand is on 12.

The short hand is on 10.

Maths ideas

In this unit you will
* talk about different times
* tell the time in hours and half-hours.

Key words

hours	clock
minutes	half past
hand	o'clock

1 Write the time shown on each clock.

a

_____ o'clock

b

_____ o'clock

c

_____ o'clock

d

_____ o'clock

e

_____ o'clock

f

_____ o'clock

2   Use colours to match the clocks and the times.

3   Draw the hour hand on each clock.

   a   6 o'clock   **b**   9 o'clock   **c**   3 o'clock

4   Look at the pictures. What time do you think it is? Fill in the hands on the clock.

Digital clocks and watches show the time in a different way.

5 What are the times on these clocks?

_____ _____ _____

6 Discuss your answers in a group. Where do you see time shown like this?

Explain

Half past the hour

This clock shows half past two.

The long hand of the clock is on the 6. This means it is **half past** an hour.

The short hand is halfway between 2 and 3. This means that it is half past 2.

7 Write the time shown on each clock.

a

half past _____

b

half past _____

c

half past _____

d

half past _____

e

half past _____

f

half past _____

8   Draw lines to match the clocks and the times.

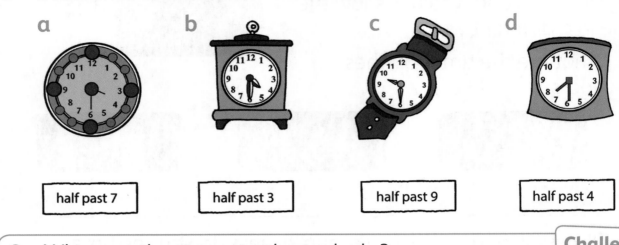

a	b	c	d
half past 7	half past 3	half past 9	half past 4

9   What are the times on these clocks?

a   **3:30**   b   **1:30**   c   **7:30**

_____   _____   _____

10   Sue started reading her book at 5 **o'clock**. She
     stopped reading at 6 o'clock. For how long did she read? _____

11   Where can you see the time in your school and community?

What did you learn?

True or false? Write T or F.

1   If the long hand of a clock is on 6, it means it is 6 o'clock.   ☐

2   This clock shows 7:30.   ☐

3   This clock shows half past six.   ☐

C Money

East Caribbean coins

The **value** of each coin is the number on the coin.

We write the values like this:
* five cents = 5¢
* one dollar = $1

The symbol **¢** means **cent**
The symbol **$** means **dollar**

5¢   10¢   25¢ $1

Maths ideas

In this unit you will
* identify **coins** in use
* count on
* add amounts using coins
* make and solve problems involving money.

Key words

value	dollar
coins	price
cent	change

Important symbols

¢ = cent
$ = dollar

1 Count each set of coins. Say how many cents altogether in each set.

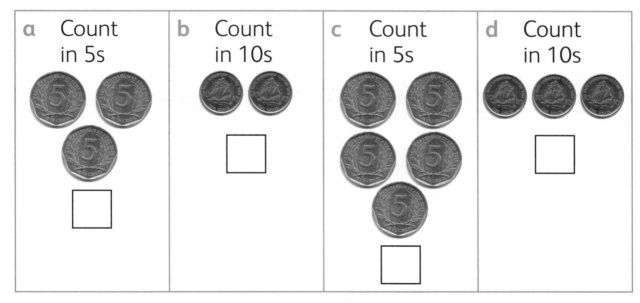

a Count in 5s	b Count in 10s	c Count in 5s	d Count in 10s

2   Count the coins. Write how much money is in each purse.

a    b    c   d

_____ c   _____ c   _____ c   _____ c

3   Count the coins. Write the amounts in order, from the smallest
     amount to the greatest.

a   b   c    d

_____   _____   _____   _____

Explain

Different coins can have the same value.

two 5¢ coins = one 10¢ coin   five 5¢ coins = one 25¢ coin

Problem-solving

4   Draw more coins in
     one hand so that
     both hands hold
     the same amount
     of money.

5   In what ways can you make 20¢? Complete the number
     sentences.

10¢ + _____ = 20¢   5¢ + _____ = 20¢   5¢ + _____ = 20¢

146

Explain

The **price** is how much we have to pay for something.
If you give too much money, you will get some back. We call the money that you get back **change**.

6 Match each price to the correct amount of money.

15¢	5 5
10¢	5 5 5 5 5
20¢	5 5 5
25¢	5 5

7 How much change will you get? Draw the coins.

Price	You pay	Change
15¢	20¢	
25¢	30¢	
30¢	40¢	

8 Make up a word problem for your group to solve. Use 5¢ and 10¢ coins. Write the problem and the answer.

_____

_____

What did you learn?

Jayden bought a bag of popcorn. The price was 15¢. He gave two 10¢ coins. How much change did he get?

Topic 10 Review

Key ideas and concepts

Days, weeks and months

* Times of day: **morning**, **midday**, **afternoon**, **evening**, **night**, **midnight**.

* There are **seven** days in one **week**.

* The days of the week are: **Sunday**, **Monday**, **Tuesday**, **Wednesday**, **Thursday**, **Friday**, **Saturday**.

* There are 12 **months** in one **year**.

* The names of the months are: **January**, **February**, **March**, **April**, **May**, **June**, **July**, **August**, **September**, **October**, **November**, **December**.

Telling the time

* A clock shows the time in hours and minutes.

* The short hand points to the hour.

* The long hand shows the minutes.

* If the long hand is on the 6, it is half past an hour.

Money

* The **value** of each coin is the number on the coin.

* The values of East Caribbean coins are: 5¢, 10¢, 25¢ and $1.

* The **price** is how much we have to pay for something.

* If you give too much money, you will get back **change**.

Think, talk, write …

1 Write the following words in your mathematics dictionary.
   Write what each word means.

   price   change   value   half past

2 Work in pairs. Make up problems for your partner to solve. Use
   real coins to work out the answers. Discuss what you have to do.

   For example: A mango costs 20¢. You have one 5¢ coin and one
   10¢ coin. Do you have enough money to buy the mango?

Quick check

1 Write the days of the week in order. Start with Monday.

2 Write the time shown on each clock.

a

b

c

d

3 Circle the correct answers.

a How many days in a week?

   7   14   30

b How many months in a year?

   28   12   7

c How much do you have if you have two 5¢ coins and one 10¢ coin?

   10¢   20¢   15¢

d A sweet costs 15¢. You give 20¢. How much change will you get?

   1¢   2¢   5¢

Teaching notes for Topic 11: Shapes and space (2)

Overview

In this topic students will focus on recognising and describing the properties of these 3-D shapes: sphere, cube, cuboid, cone, cylinder.

A Solid shapes around us	* Three-dimensional or 3-D shapes are solid shapes. The three dimensions are length, depth and width (breadth). * Help students to understand that solid shapes take up air space. Our bodies are 3-D shapes, for example. * Give the students real objects to examine and let them note how the shapes are the same or different. Help them to look at the different faces (sides) and the corners of a shape.
B Describing and sorting solid shapes	* Recognising shapes helps students to sort and classify, which is an important concept in mathematics. The students need to be able to describe and sort the following solid shapes: * A cube has six square faces that are all the same size. (A cube also has eight corners and twelve edges.) * A sphere has a curved face, with no edges or corners. * A cuboid has six faces; the four rectangular faces are the same and the two faces at the ends are the same. * A cylinder has a curved surface and two circular faces at the ends. * A cone has a circular face at one end (the base) and a point. The other surface is curved.
C Exploring solid shapes	* 3-D shapes can slide, roll and stack. * Use real (unbreakable) objects to teach students about the properties of rolling, stacking and sliding.

Mathematical vocabulary

Students need to learn vocabulary that they can use to describe 3-D shapes, and also learn to differentiate between them, for example: **faces**, **sides**, **corners**, **curved** and **flat**. They also need to learn about the properties: **roll**, **stack** and **slide**.

Notes for home

Your child is learning about the properties of 3-D objects. Help them understand that these objects can roll, slide and stack. Give them a selection of objects that will not break and let them find out what each object can do. For example, they can find out that a clean, empty plastic bottle can slide and roll, empty boxes can slide and stack, toilet rolls can roll and slide and stack.

Topic 11 Shapes and space (2)

 A **Solid shapes around us** *(pages 154–156)*

How many different shapes can you see in these pictures?

What objects around you have the same shape as a ball?

Describing and sorting solid shapes
(pages 157–160)

Can you find five objects in your classroom that are shaped like a box? Describe the shape.

Describe the shape of the ice-cream cone. Why is it this shape? Can you think of a different shape for an ice-cream cone?

C Exploring solid shapes *(pages 161–163)*

What will happen to the logs on this truck if you take away the supporting poles on the sides? How do you know?

A Level 1 class made a building using different shapes. What shapes did they use? Can you use the same shapes to make something different?

A Solid shapes around us

Explain

These are **solid** shapes.

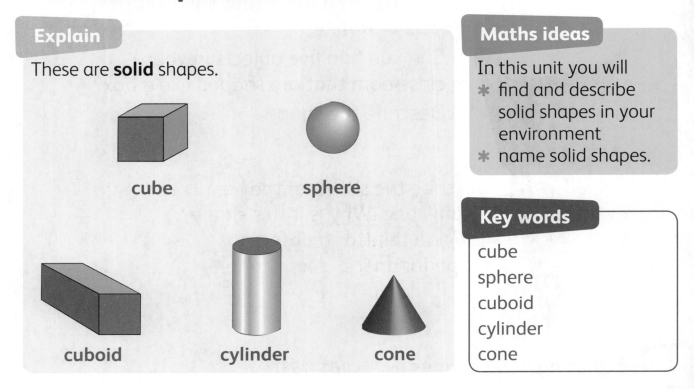

cube

sphere

cuboid

cylinder

cone

Maths ideas

In this unit you will
* find and describe solid shapes in your environment
* name solid shapes.

Key words

cube
sphere
cuboid
cylinder
cone

1 Circle the shapes that are the same in each row.

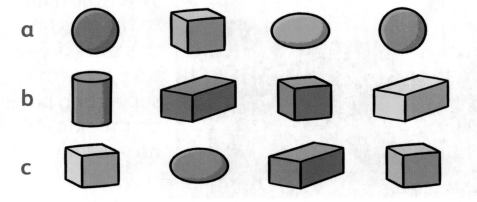

a

b

c

2 Underline the name of each shape.

a

sphere cube cuboid

b

sphere cube cuboid

c

sphere   cube   cuboid

d

cylinder   cone   cube

e

cylinder   sphere   cuboid

f

cylinder   sphere   cone

3   This is a toy store. Colour the **cubes red**. Colour the **cuboids blue**.
Colour the **spheres green**.

Investigate

4 Find three cuboids in your classroom. Draw them and then label your drawings.

5 Which shape do you see least often? Share your ideas.

What did you learn?

Complete the sentence.

Cones, _____, _____,

_____ and _____

are _____ shapes.

B Describing and sorting solid shapes

Explain

Cubes and cuboids have six flat faces.

face

face

This is a cube.   This is a cuboid.

The six faces of a cube are square. They are all the same size.

A sphere is curved and has no flat faces.

face

face

A cylinder has two round flat faces. The other parts are curved.

A cone has just one round flat face.

Maths ideas

In this unit you will
* describe and compare solid shapes
* sort shapes.

Key words

cubes
cuboids
faces
sphere
curved
flat
cylinder
cone

1   Look around your classroom.

   a   Which shape can you find the most of?   _____

   b   Draw two things that are that shape.

2 **a** Sort the following into three groups.

b Draw your groups here. Give each group a name to show how you sorted the items in the group.

Group 1:
Group 2:
Group 3:

3 Sort the following shapes into four groups. Use colours to show the four different groups.

4 Find an example of a solid shape at home or at school. Draw the shape and then label it. Explain to your partner why it is a solid shape.

5 Work out the shapes in these riddles.

   a It is a solid shape. It has one round flat face. One end has a point.

   _____

   b It is a solid shape. It has six faces that are all the same.

   _____

6 Make up a riddle about a solid shape. Let your partner answer.

Riddle: _____

_____

Answer: _____

Investigate

7 Find out what this solid shape is called.

8 Where can you see shapes like these?

What did you learn?

1 Compare a cube and a cuboid.

   a What is the same?

   b What is different?

2 Compare a cylinder and a cone.

   a What is the same?

   b What is different?

C Exploring solid shapes

Explain

Solid shapes can move in certain ways because of their shapes.

These shapes can **stack**.

These shapes can **roll**.

These shapes can **slide**.

Maths ideas

In this unit you will
* learn which shapes can roll, slide and stack
* make objects with solid shapes.

Key words

stack
roll
slide

Think and talk

Can you see why only some shapes can stack?

1 Look at the shapes.

a Can the yellow shape roll? Why?

_____ because _____

b Does the green shape stack?

_____ because _____

c Can you slide the red shape?

_____ because _____

d Which shapes cannot roll?

_____

e Which shapes cannot stack?

_____

f Which shapes can stack and roll?

_____

2 Look at these shapes and then answer the questions.

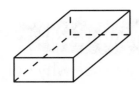

a Colour all the shapes that can roll.

b Circle the shapes that can stack.

c Tick the shapes that can slide.

3 Look at the shapes shown in Question 2, and then tick ✓ the columns to complete the table.

Shape	Can it roll?	Can it stack?	Can it slide?

Shape	Can it roll?	Can it stack?	Can it slide?
(triangular pyramid)			
(cylinder)			
(cone)			

a   Which shape can roll, stack and slide? _____

b   Which shape can roll, but not slide? _____

c   Which shapes cannot stack? _____

Investigate

4   Use modelling clay to make a shape that will roll, but not in a straight line.
What shape did you make? [_____]

5   Make a model of a building using as many different solid shapes as you can. Try to use cubes, cuboids, cylinders and cones.

6   Draw a picture of your model or take a photograph and paste it here.

What did you learn?

True or false? Write T or F.

1   Cuboids and cylinders can slide. [ ]

2   Cubes and cones can roll. [ ]

3   Spheres can stack. [ ]

4   Cubes can stack and slide. [ ]

Topic 11 Review

Key ideas and concepts

* These are solid shapes: cube, sphere, cuboid, cylinder, cone.
* Some shapes are curved or round.
* Some shapes have flat faces.
* Solid shapes can slide, roll and stack depending on their shape.

Think, talk, write …

1   Write the names of these shapes in your mathematics dictionary under the heading **Solid shapes**. Write one sentence to describe each shape. The first one has been done for you.

   a cube: This has six flat faces that are all the same size.

   b cuboid   **c** sphere

   d cylinder   **e** cone

2   Take two objects from your desk and move them. Then complete the sentences to describe how they move.

   The [＿＿＿] can [＿＿＿] and [＿＿＿].

   The [＿＿＿] can [＿＿＿], but it can't [＿＿＿].

Quick check

1   Match the shapes and the names.

| cone | sphere | cuboid | cube | cylinder |

2   What can these shapes do? Underline the correct words.

   a cube: stack   roll   slide   **b** sphere: stack   roll   slide

   c cylinder: stack   roll   slide   **d** cone: stack   roll   slide

Teaching notes for Topic 12: Measurement (3)

Overview

In this topic students will develop further skills in estimating and taking accurate measurements. They will measure mass, capacity and temperature.

A Mass	* Mass is the amount of matter in an object. It tells us how heavy something seems. Weight is not the same as mass, but we often talk about how much something weighs in everyday life, when what we are really talking about is its mass.
	* Teach the students to meaure mass in kilograms. The kilogram is a metric unit of mass. One kilogram is about as heavy as a litre of water or juice.
	* You can use a balance scale with weights to measure mass. You can use stones or metal weights on the scale.
	* Students need to develop their skills of estimating as well to do quick measurements and check real measurements. Start by letting students think about mass and containers that they know and let them compare these with things they want to measure.
B Capacity	* Capacity tells us how much a container can hold.
	* Students first learn to use containers such as buckets, cups and plastic bottles to measure capacity.
	* Then teach the students to measure capacity in litres. Many cold drinks are sold in litre bottles. You can also use measuring jugs or cylinders.
C Temperature	* Temperature tells us how hot or cold something is.
	* Teach the students to describe temperature by using vocabulary such as hot, cold and cool. They will learn to measure accurately in degrees Celsius and Fahrenheit later.

Mathematical vocabulary

Students need to learn to use vocabulary such as **kilogram**, **litre**, **more than**, **less than**, **as … as**, **mass**, **capacity** and **temperature**. They also need to understand the concept of **estimating**.

Notes for home

Your child is learning how to estimate mass and capacity using non-standard units and then how to take accurate measurements in kilograms and litres.

Let your child take simple measurements at home. Your child can estimate and then measure which containers hold more, or less water. Let them fill up two or three different containers and then measure the water in each container by pouring it into an empty 1 litre milk or cold drink bottle.

Children can also estimate their own mass and then weigh themselves on a scale. (We often talk about how much things 'weigh' in everyday life when we mean mass, although mass and weight are not the same.)

12 Measurement (3)

A Mass (pages 168–171)

> Which of these can you pick up easily? Why can you not pick up the potted plant?

Discuss the pictures.

* Which objects are heavier than the girl? Colour those objects blue.

* Which objects are lighter than the girl? Colour those red.

Capacity *(pages 172–175)*

What would be the quickest way to fill this bucket with water?

Use the cup? Use the jug? Why?

C

Is the water hot or cold? How do you know? Why do you need to be careful?

Temperature *(pages 176–179)*

Do these pictures show hot or cold temperatures?

A Mass

Some objects are **heavy**.
Some objects are **light**.
A rock is **heavier than** a feather.
A feather is **lighter than** a rock.
We can **weigh** things to find out how heavy they are.

Maths ideas

In this unit you will
* estimate how heavy objects are
* measure how heavy objects are in **kilograms** (kg)
* compare objects.

Key words

heavy
light
heavier than
lighter than
weigh
kilograms

1   Look at the pictures.
    Circle the object that is
    heavier in each pair.

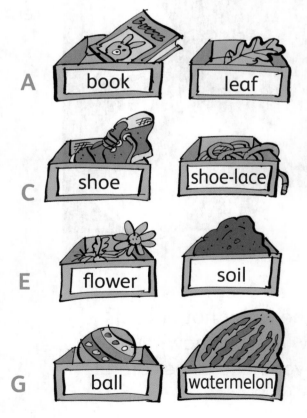

2   Arrange these objects in order from the lightest to the heaviest.

a feather
a stone
a person
a car
a computer
a ball

_____ ➔ _____ ➔ _____ ➔

_____ ➔ _____ ➔ _____

Investigate

3   Work in pairs. Collect 3 or 4 plastic bags or cardboard boxes.
Discuss which bag or box is the heaviest.
How can you measure this? Share your ideas with the class.

Explain

We can use a balance scale to compare the mass of different objects.

These tomatoes have the same
mass as one bag of beans.
The scale is balanced.

These bananas have a mass that
is more than the bag of beans.
The heavier pan drops down.

4 How can you measure the mass of these objects? First estimate, and then measure. Say what you used to get your measurements.

a

b

Estimate: _____

Measurement: _____

Estimate: _____

Measurement: _____

c

d

Estimate: _____

Measurement: _____

Estimate: _____

Measurement: _____

Explain

When we weigh an object, we find the **mass** of the object.
We measure mass in **kilograms**.

These sweet potatoes have a mass of one kilogram (1 kg).

These crisps have a mass that is less than 1 kg.

These bricks have a mass that is more than 1 kg.

5 Look at the following pictures. Estimate.
Which objects have a mass of more than 1 kg?
Which objects have a mass of less than 1 kg?

Draw the objects or write their names in the correct column.

more than 1 kg	less than 1 kg

What did you learn?

1 Write down three things that are heavier than you.

2 Write down two things that are lighter than 1 kg.

B Capacity

Explain

Capacity tells us how much liquid a **container** can hold.

This glass is **empty**.   This glass is **full**.

These containers can all hold liquids such as tea, water or juice.

The cup holds **less than** the jug.
The bucket holds **more than** the jug.

Maths ideas

In this unit you will
* estimate how much liquid containers can hold
* measure liquid in **litres** (ℓ)
* compare containers.

Key words

container
empty
full
litres
less than
more than

1   Colour in the liquid in the containers.
    Then label each picture full or empty.

_____   _____   _____   _____   _____

2   Which container can hold the most liquid? Circle the container.

a

b

Explain

We can estimate how much a container can hold.
We can measure the **amount** of water that a container can hold in cups.

This water bottle can hold
about 4 cups of water.

This bottle can hold
about 8 cups of water.

Investigate

3   Work in pairs. How can you measure
    how much water there is in the bucket?
    Discuss with your partner how you
    measured this.

4   Your teacher will give you a jug and a cup. How many cups of
    water will you need to fill the jug? Write down what you have
    estimated. Then check your estimate by filling the jug using
    the cup. Count the cups as you fill the jug.

    Estimate: ☐   How many cups did you count? ☐

5 Look at the pictures. Estimate how much each container can hold. Circle the best answer.

a

The bowl can hold 4 cups / 24 cups of water.

b

The bath can hold 10 buckets / 2 buckets of water.

6 Which units can we use to measure these? Circle the correct units.

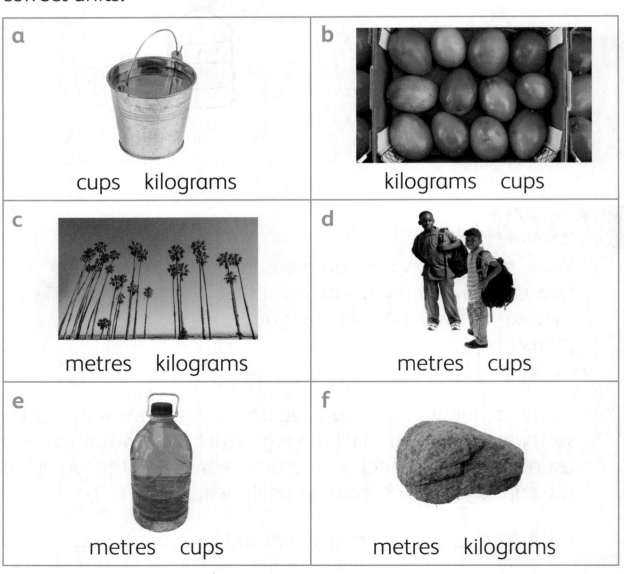

a	b
cups kilograms	kilograms cups
c	**d**
metres kilograms	metres cups
e	**f**
metres cups	metres kilograms

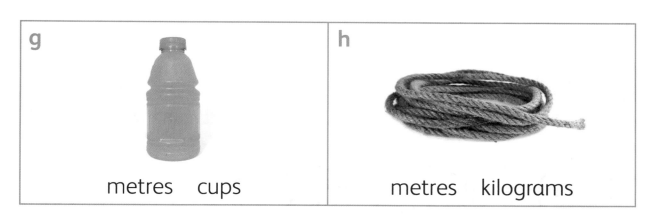

g	h
metres cups	metres kilograms

Problem-solving

7 Pablo has to fill a bucket with water. He has a
plastic cup and a plastic jug.
Which container should he use to fill up the bucket? _____
Why? _____

8 A jug holds five cups of juice. How many cups of juice in
three jugs?

_____ + _____ + _____ = _____

Challenge

9 You have a bucket of water and two empty buckets.
You have to pour water from the full bucket into
the empty buckets. You want the same amount of water in
each bucket.

Discuss how you can do this.

What did you learn?

True or false? Write T or F.

1 A cup holds more water than a jug. ☐

2 We can measure how much liquid a container holds. ☐

C Temperature

Explain

These things are **hot**.

These things are **cold**.

Hot and cold describe the **temperature**.

Maths ideas

In this unit you will
* describe the temperature of an object
* use words such as hot and cold.

Key words

hot
cold
temperature

1 Look at the pictures. Write **hot** or **cold** under each picture.

_____ _____

2 Draw a picture of the clothes you wear when it is cold. Label the items of clothing in your picture.

3 Is it hot or cold? Write the letters in the correct place in the table.

Hot	Cold

4 Look at the picture and then answer the questions.

   a Is it a hot day or a cold day?

      _____

   b Circle in red three hot things in the picture.

   c Circle in blue three cold things in the picture.

Investigate

5 Make a list of three things that are too hot to touch.

Make a poster to warn people about these things. You can draw pictures or write words. Use the frame on the next page to draw your poster.

6 How can you make these things warm? Give one idea for each.

   a food:

   b your house:

What did you learn?

Circle the things that are hot.

Topic 12 Review

Key ideas and concepts

* We weigh an object to find the **mass** of the object.
* Some objects are **heavier** or **lighter than** other objects.
* We measure mass in **kilograms**.
* We can measure the **amount** of liquid that a container can hold.
* Some containers hold **more** or **less** liquid than other containers.
* The words **hot** and **cold** describe the **temperature**.

Think, talk, write ...

1   Match these words and definitions. Copy the words into your mathematics dictionary.

mass	We use this to measure mass.
temperature	A measure of the amount that a container can hold.
kilogram	How hot or cold something is.
capacity	How heavy or light an object is.

2   Explain to your partner how you can find out how much water a container can hold.

Quick check

1   Circle the object that is heavier.

a   b

2   Circle the container that holds more.

a   b

Teaching notes for Topic 13: Data handling (2)

Overview

In this topic students will learn how to read and interpret simple block graphs and bar graphs. Then they will start to learn how to collect data in order to make their own block and bar graphs.

A Block and bar graphs	* Make sure that students know what data is and that they remember what a picture graph is.
	* Teach the students about block graphs and bar graphs. They need to be able to interpret data shown on both types of graphs. You do not need to teach them the differences between the two types.
	* Block and bar graphs represent data. They show us the values of data (how much and how many, for example) in vertical or horizontal blocks or strips. The blocks help us to compare the value of the data in a visual way.
	* Bar graphs need titles and labels in order to make sense. Teach the students to read the titles and labels on graphs.
B Collecting data and making block and bar graphs	* Build on what the students already know about collecting data. Provide as many practical activities as you can so that students get used to recording information accurately.
	* Help students to make their own very simple graphs with the data they have collected. Provide grid paper for this purpose. Teach the students to label the graphs and to give them titles. Write these on the board for the students to copy.

Mathematical vocabulary

Students need to be able to understand the following words: **picture graph (pictograph)**, **block graph**, **bar graph**, **key**, **data**, **information**, **tally** and **symbol**.

Notes for home

The children are using what they already know about collecting and sorting data to make and draw graphs. You can help your child by making simple graphs at home. For example, put a chart on the fridge and let family members colour blocks to show how many glasses of water they drink each day. Or you could make picture graphs to show how many different types of birds you see in the garden during the course of a week.

A Block and bar graphs *(pages 184–188)*

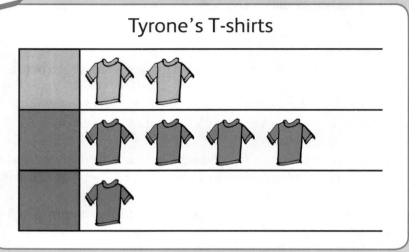

Tyrone's T-shirts

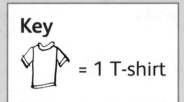

Key

= 1 T-shirt

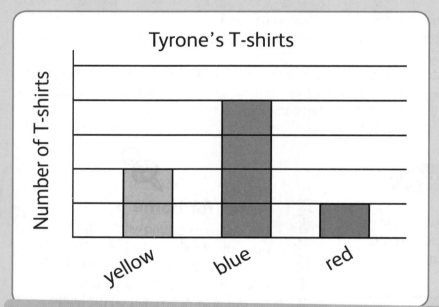

Tyrone's T-shirts

These are both graphs and they show the same data. What is different about them? Which one is easier to understand? What do you think is Tyrone's favourite colour?

Look at the graphs again.

* Why do you think the graphs have headings?
* How do you know how many blue T-shirts are shown on the block graph?

Collecting data and making block and bar graphs *(pages 189–193)*

How do you think the ice-cream shop decides which flavours to make?

A teacher collected data from her class. She asked each learner this question:

What is your favourite subject?

She recorded these answers:

* Mathematics: 11
* Reading: 10
* Art: 9

How could she make a graph with this information? Discuss and complete the graph.

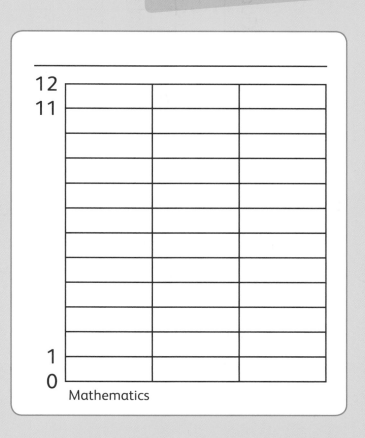

A Block and bar graphs

Maths ideas

In this unit you will
* read block graphs
* compare information in graphs.

Explain

We can make **block graphs** to show **data**.

The **title** and **labels** on the graph tell us what the data is.

Look at the graph below.
* The title tells us that the graph shows the number of goals scored by football players.
* Each block on the graph represents one goal.
* The labels tell us the names of the football players and the number of goals they scored.

Key words

block graphs
data
title
labels

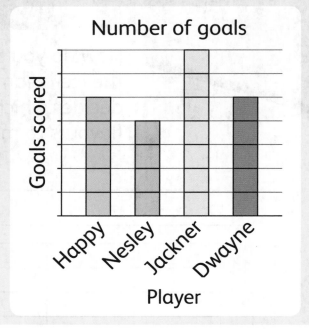

1   Look at the graph about the goals scored. Answer the questions.

   a   Which player scored the most goals? _____

   b   Which players scored the same number of goals?
       _____ and _____

   c   How many goals did Nesley score? _____

   d   How many more goals did Jackner score than Dwayne?
       _____ – _____ = _____

2 a Which food is the most popular?

_____

b How many people like cracked conch?

_____

c How many people liked Johnnycakes or souse?

_____ + _____ =

_____

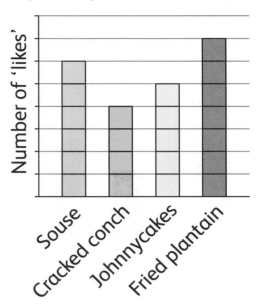

My family's favourite foods

Number of 'likes'

Souse Cracked conch Johnnycakes Fried plantain

3 A health worker drew this graph.

a What data does this graph show?

_____

b How many students had measles?

_____

c How many students had tonsillitis?

_____

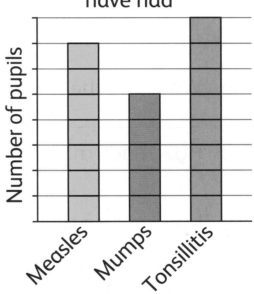

Illnesses that pupils have had

Number of pupils

Measles Mumps Tonsillitis

4 Bianca found out how many books her friends read during the holidays. Answer the questions about the graph that Bianca drew.

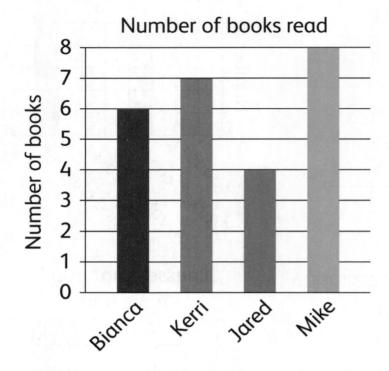

Number of books read

> **Explain**
>
> This is a bar graph. The number of books is shown by the length of the bar and the numbers on the scale.

a Who read the fewest books?

_____

b How many more books did Mike read than Jared?

_____ − _____ = _____

c How many books did Bianca and Kerri read altogether?

_____ + _____ = _____

d Work out how many books the four children read altogether.

_____ + _____ + _____ + _____ = _____

5 The following graphs show the number of people who went surfing at two different beaches over three days.

Compare the data on the graphs.

a How many people surfed at Long Beach on Friday? _____

b Were there more surfers at Long Beach or Pebble Bay on Saturday? _____

c On which day were there the most surfers at both beaches?

_____

d How many people surfed altogether on Friday?

_____ + _____ = _____

e Which beach was the most popular on Sunday?

_____

f How many people surfed at Pebble Bay on Friday and Sunday?

_____ + _____ = _____

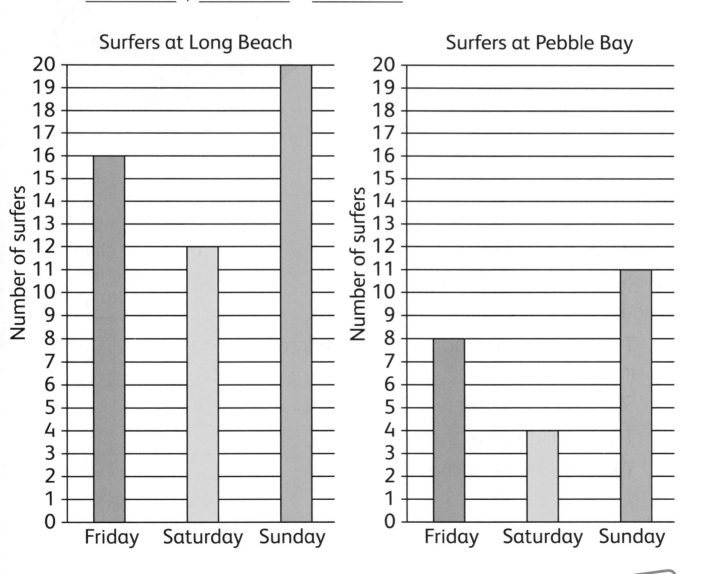

6 Seven students had to choose a sandwich they liked. Look at the graph. Work out how many learners did NOT choose a fish sandwich.

_____

Sandwiches you can choose from

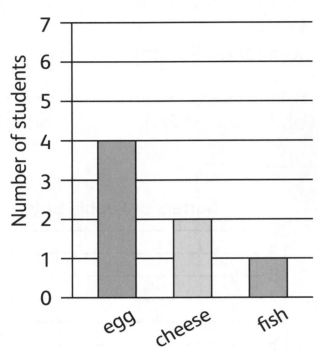

What did you learn?

What data does this graph show?
Underline the sentences that are true.

1 The graph shows data about children in a group.

2 Six children do not wear spectacles.

3 Three children wear spectacles.

4 Ten children wear spectacles.

How many children wear spectacles in my group

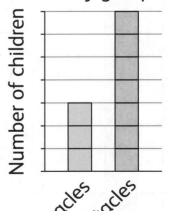

188

B Collecting data and making block and bar graphs

Explain

In Topic 9 you learnt:
* how to collect data by asking questions and making observations
* how to record data by using tallies.

Maths ideas

In this unit you will
* collect **data**
* use blocks to show data on a **block graph**.

1 a Colour the beads using one colour for each shape of bead. Use red, blue, yellow and green.

Key words

data

block graph

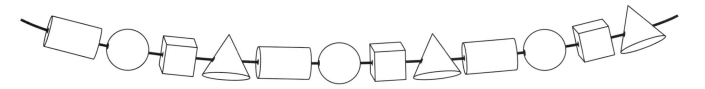

b Complete the block graph.

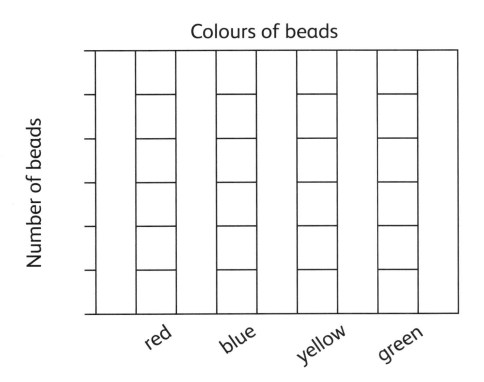

Colours of beads

Number of beads

red   blue   yellow   green

c Fill in the table to show how many there are of each colour.

Colour	Number
red	
blue	
yellow	
green	

2 Find out which of these sports your friends like to play.
Ask 15 people.
Each person gives one answer.
Use tallies to record the answers.

What sports do you like to play?

football	
netball	
swimming	
running	

3 Make a block graph to show the information you collected in Question 2. Colour the blocks for each sport in a different colour.
Remember: one block = one answer

Number of students

football   netball   swimming   running

Problem-solving

4 Mel eats three plums. Peter eats six plums. Dwayne eats seven plums.

Who recorded this information correctly, A or B? ________

A

Mel	///
Peter	𝍸 /
Dwayne	𝍸 //

B

Mel	𝍸 𝍸 𝍸
Peter	𝍸 /
Dwayne	𝍸 ///

5 Can you use a bar graph to solve problems? Read the graph and answer the questions.

Mrs Pinder runs a guesthouse. She serves five different kinds of fruit at breakfast. She wants to cut down and only serve three different fruits.

a What does the graph show? ____________________

b Which fruits should she cut out? ____________________

and ____________________

c Why? ____________________

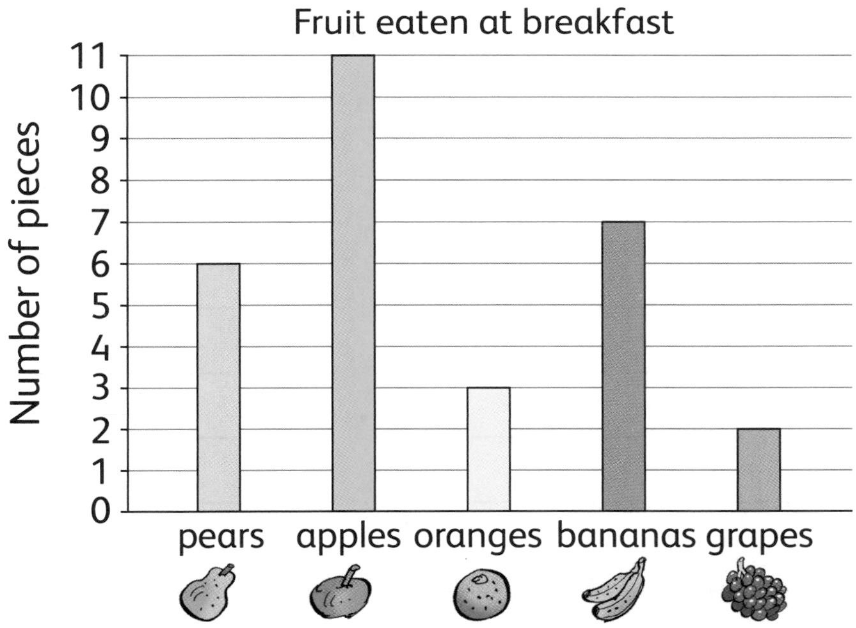
Fruit eaten at breakfast
Number of pieces
0
1
2
3
4
5
6
7
8
9
10
11
pears
apples
oranges
bananas
grapes

Problem-solving

6 Your school wants to offer two new sports for the pupils to play. How can you find out which sports to add? Make some suggestions.

_____

_____

_____

Collect data and record it here.

Title: _____

Draw a block graph to show your data.

_____ _____ _____ _____ _____

7 Mrs Brown-Simmonds has a shoe shop. She wanted to find out what colour shoes her customers liked. She asked 16 people.

Four people like red shoes and four other people liked white shoes. Six people liked black shoes. Only two people liked grey shoes.

Draw a graph to show this data.

What did you learn?

1 Name two types of graph that you can use to show data.

2 Complete the sentence.

A graph must have a [   ] and

[   ] to tell us what the graph is about.

Topic 13 Review

Key ideas and concepts

* We can make **block** or **bar graphs** to show **data**.
* The **title** and **labels** on the graph tell us what the data is.
* We need to **collect** and **record** data to use on graphs.

Think, talk, write ...

1 Find three new mathematics words that you have learnt in this topic. Write the words in your mathematics dictionary. Write a definition of each word or draw a picture to show what it means.

2 Discuss why graphs can be useful.

Quick check

1 Answer the questions about the graph.

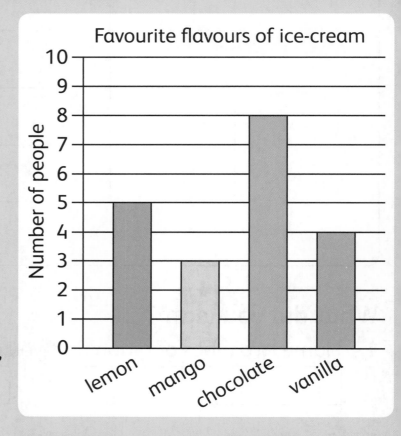

Favourite flavours of ice-cream

a How many people like chocolate?

b How many people like lemon?

c Which is more popular, mango or vanilla?

d How many people answered the question about their favourite ice-cream flavours?

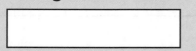

□ + □ + □ + □ = □

Test yourself (3)

1   Look at the calendar. Complete the sentences.

March

Sunday	Monday	Tuesday	Wednesday	Thursday	Friday	Saturday
		1	2	3	4	5
6	7	8	9	10	11	12
13	14	15	16	17	18	19
20	21	22	23	24	25	26
27	28	29	30	31		

a   The month is _____.

b   This month has _____ days.

c   The month after this is _____.

d   There are _____ Mondays in this month.

e   The first day of the month is on a _____.

f   The last day of the month is on a _____.

2   Look at the bar graph.
    The graph shows at what
    time some people go to
    bed at night.

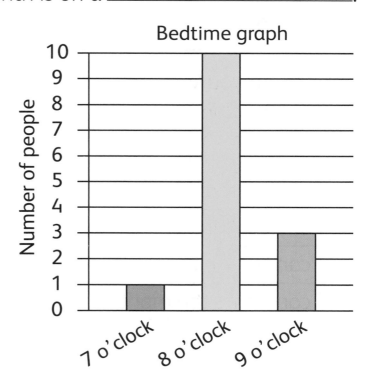

Bedtime graph

a How many people go to bed at 7 o'clock? _____

b How many people go to bed at 9 o'clock? _____

c At what time do most people go to bed? _____

d How many people answered the question to collect this data?

_____ + _____ + _____ = _____

3 Match the names with the correct shapes.

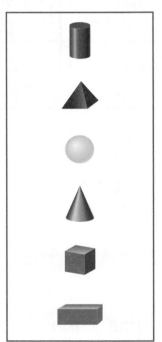

cube

cylinder

cuboid

sphere

pyramid

cone

4 What do you know?
Tick the column to show your answer.

	Yes	No	Not always
I can estimate mass.			
I can explain what capacity is.			
I can tell the time.			
I can name the months of the year.			
I can name the days of the week.			